CAMBRIDGE LIBRARY COLLECTION

Books of enduring scholarly value

Perspectives from the Royal Asiatic Society

A long-standing European fascination with Asia, from the Middle East to China and Japan, came more sharply into focus during the early modern period, as voyages of exploration gave rise to commercial enterprises such as the East India companies, and their attendant colonial activities. This series is a collaborative venture between the Cambridge Library Collection and the Royal Asiatic Society of Great Britain and Ireland, founded in 1823. The series reissues works from the Royal Asiatic Society's extensive library of rare books and sponsored publications that shed light on eighteenth- and nineteenth-century European responses to the cultures of the Middle East and Asia. The selection covers Asian languages, literature, religions, philosophy, historiography, law, mathematics and science, as studied and translated by Europeans and presented for Western readers.

Bija Ganita, or, the Algebra of the Hindus

An important mathematician and astronomer in medieval India, Bhascara Acharya (1114–85) wrote treatises on arithmetic, algebra, geometry and astronomy. He is also believed to have been head of the astronomical observatory at Ujjain, which was the leading centre of mathematical sciences in India. Forming part of his Sanskrit magnum opus *Siddhānta Shiromani*, the present work is his treatise on algebra. It was first published in English in 1813 after being translated from a Persian text by the East India Company civil servant Edward Strachey (1774–1832). The topics covered include operations involving positive and negative numbers, surds and zero, as well as algebraic, simultaneous and indeterminate equations. Strachey also appends useful notes made by the orientalist Samuel Davis (1760–1819). Of enduring interest in the history of mathematics, this was notably the first work to acknowledge that a positive number has two square roots.

Cambridge University Press has long been a pioneer in the reissuing of out-of-print titles from its own backlist, producing digital reprints of books that are still sought after by scholars and students but could not be reprinted economically using traditional technology. The Cambridge Library Collection extends this activity to a wider range of books which are still of importance to researchers and professionals, either for the source material they contain, or as landmarks in the history of their academic discipline.

Drawing from the world-renowned collections in the Cambridge University Library and other partner libraries, and guided by the advice of experts in each subject area, Cambridge University Press is using state-of-the-art scanning machines in its own Printing House to capture the content of each book selected for inclusion. The files are processed to give a consistently clear, crisp image, and the books finished to the high quality standard for which the Press is recognised around the world. The latest print-on-demand technology ensures that the books will remain available indefinitely, and that orders for single or multiple copies can quickly be supplied.

The Cambridge Library Collection brings back to life books of enduring scholarly value (including out-of-copyright works originally issued by other publishers) across a wide range of disciplines in the humanities and social sciences and in science and technology.

Bija Ganita
or, the Algebra
of the Hindus

BHASCARA ACHARYA
TRANSLATED BY EDWARD STRACHEY

CAMBRIDGE
UNIVERSITY PRESS

CAMBRIDGE UNIVERSITY PRESS

Cambridge, New York, Melbourne, Madrid, Cape Town,
Singapore, São Paolo, Delhi, Mexico City

Published in the United States of America by Cambridge University Press, New York

www.cambridge.org
Information on this title: www.cambridge.org/9781108056014

© in this compilation Cambridge University Press 2013

This edition first published 1813
This digitally printed version 2013

ISBN 978-1-108-05601-4 Paperback

BIJA GANITA,

&c. &c. &c.

BIJA GANITA:

OR THE

ALGEBRA

OF THE

HINDUS.

BY

EDWARD STRACHEY,

OF THE

EAST INDIA COMPANY's BENGAL CIVIL SERVICE.

LONDON:

PRINTED AND SOLD BY W. GLENDINNING, 25, HATTON GARDEN.

1813.

CONTENTS.

CONTENTS.

CONTENTS.

MR. DAVIS'S NOTES.

ERRATA.

Page   3, line   4, *for* Heilbronnen *read* Heilbronner.

        5,   1, *for* undistinguished *read* undistinguishing.

     11,   6, *after* notes *read* at the bottom of the pages.

     13,   1, *dele* the inverted commas at the beginning of the line.

     17,   5, (at the end), *for* number *read* colour.

     29,   25, Suppose $\dfrac{ax + c}{b} = y$ where a, b, and c are known and x and y unknown.

     93,   9, 10, and 11, the character prefixed to the numbers 1 and 2 is here लो which is the first letter of the word *Loheet* (see opposite page); but it should be a different character, viz. the first letter of the word *Roop* रू.

Note—In Mr. Davis's notes the word Ja, Roo, Bha, Ca, &c. which are frequently used, are contractions of Jabut, Roop, Bhady, Canist, &c. They should have been printed with points after them, thus, Ja. Roo. &c.

PREFACE.

It is known that there are Sanscrit books on Astronomy and Mathematics. Whether the Science they contain is of Hindoo Origin and of high antiquity, or is modern and borrowed from foreign sources, is a question which has been disputed. Some of the Advocates for the Hindoos have asserted their pretensions with a degree of zeal which may be termed extravagant; and others among their opponents have with equal vehemence pronounced them to be impostors, plagiaries, rogues, blind slaves, ignorant, &c. &c.

My object in the following paper is to support the opinion that the Hindoos had an original fund of Science not borrowed from foreign sources. I mean to infer also, because of the connexion of the sciences and their ordinary course of advancement, that the Hindoos had other knowledge besides what is established by direct proof to be theirs, and that much of what they had, must have existed in early times.

But with respect to the antiquity of the specimens which I am going to exhibit, nothing seems to be certainly known beyond this, that in form and substance as they are here, they did exist at the end of the 12th or the beginning of the 13th Century.

It is not my purpose to inquire here what parts of Indian science have already been ascertained to be genuine. I only wish to observe that the doubts which have been raised as to the pretensions of the Hindoos are of very recent birth, and that no such doubts have been expressed by persons who were perhaps as well able to judge of the matter as we are.*

---

* The Edinburgh Review, in criticising Mr. Bentley's Indian Astronomy, in the 20th number, ably contended for the antiquity and originality of Hindoo Science. The writer of that article however seems to have left the field; and his successor, in a Review of Delambre's History of Greek Arithmetic, has taken the other side of the question, with much zeal. This Critic is understood to be Mr. Leslie, who, in his Elements of Geometry, has again attacked the Hindoos. Mr. Leslie, after explaining the rule for constructing the sines by differences, which was given in the 2nd Volume of the Asiatic Researches by Mr. Davis, from the Surya Siddhanta, adds the follownig remarks.

We are told that in early times Pythagoras and Democritus, who taught the Greeks astronomy and mathematics, learnt these sciences in India. The Arabians

---

" Such is the detailed explication of that very ingenious mode which, in certain cases, the Hindoo Astro-
" nomers employ for constructing the table of approximate sines. But totally ignorant of the principles of
" the operation, those humble calculators are content to follow blindly a slavish routine. The Brahmins must
" therefore have derived such information from people farther advanced than themselves in science, and of
" a bolder and more inventive genius. Whatever may be the pretensions of that passive race, their know-
" ledge of trigonometrical computation has no solid claim to any high antiquity. It was probably, before
" the revival of letters in Europe, carried to the East, by the tide of victory. The natives of Hindustan
" might receive instruction from the Persian Astronomers, who were themselves taught by the Greeks of
" Constantinople, and stimulated to those scientific pursuits by the skill and liberality of their Arabian con-
" querors."—(Leslie's Elements, p. 485.)

When scientific operations are detailed, and most of the theorems on which they depend are given in the form of rules, surely it is not to be inferred because the demonstrations do not always accompany the rules, therefore that they were not known; on the contrary, the presumption in such a case is that they were known. So it is here, for the Hindoos certainly had at least as much trigonometry as is assumed by Mr. Leslie to be the foundation of their rule. Mr. Leslie, after inferring that the Hindoos must have derived their science from people farther advanced than themselves, proceeds to shew the sources from which they might have borrowed, namely, the Persians, the Greeks, and the Arabians. Now as for the Persians as a nation, we do not know of any science of theirs except what was originally Greek or Arabian. This indeed Mr. Leslie would seem not to deny; and as for the Greeks and Arabians it is enough to say that the Hindoos could not borrow from them what they never had. They could not have borrowed from them this *slavish routine* for the sines, which depends on a principle not known even to the modern Europeans till 200 years ago. In short the tide of victory could not have carried that which did not exist.

It appears from Mr. Davis's paper that the Hindoos knew the distinctions of sines, cosines, and versed sines. They knew that the difference of the radius and the cosine is equal to the versed sine; that in a right-angled triangle if the hypothenuse be radius the sides are sines and cosines. They assumed a small arc of a circle as equal to its sine. They constructed on true principles a table of sines, by adding the first and second differences.

From the Bija Ganita it will appear that they knew the chief properties of right-angled and similar triangles.

In Fyzee's Lilavati I find the following rules:

(The hypothenuse of a right-angled triangle being h, the base b, and the side s.)

Assume any large number p, then $\dfrac{\sqrt{((b^2+s^2)\,p^2)}}{p} = h.$

$$b = \sqrt{(h^2-s^2)} \text{ and } s = \sqrt{(h^2-b^2)}.$$
$$\sqrt{((b-s)^2+2bs)} = h.$$
$$(h+b)(h-b) = h^2-b^2.$$

b being given to find h and s in any number of ways; let p be any number; then $\dfrac{2pb}{p^2-1} = s$, and $ps-b = h.$

$$\frac{\dfrac{b^2}{p}-p}{2} = s, \text{ and } \frac{\dfrac{b^2}{p}+p}{2} = h.$$

h being given, $\dfrac{2ph}{p^2+1} = s$, and $ps-h = b.$

always considered the Indian astrology and astronomy as different from theirs and the Greeks. We hear of Indian astronomy known to them in the time of the Caliph Al Mamun. (See d'Herbelot). Aben Asra is said to have compared the Indian sphere with the Greek and Persian spheres. (Heilbronnen Hist. Math. p. 456). We know that the Arabians ascribe their numeral figures to the

---

Let p and q be any numbers; then

$$2pq = s, \quad p^2 - q^2 = b, \text{ and } p^2 + q^2 = h.$$

Given $a = h \pm s$; then $\dfrac{a - \frac{b^2}{a}}{2} = s$, and $\dfrac{a + \frac{b^2}{a}}{2} = h.$

Given $a = b + s$; then $\dfrac{a - \sqrt{(2h^2 - a^2)}}{2} = b$, and $\dfrac{a + \sqrt{(2h^2 - a^2)}}{2} = s.$

There are also rules for finding the areas of triangles, and four-sided figures; among others the rule for the area of a triangle, without finding the perpendicular.

For the circle there are these rules (c being the circumference, D the diameter, c the chord, v the versed sine, a the arch,)

$$\text{c} : \text{D} :: 22 : 7; \text{ and c} : \text{D} :: 3927 : 1250. \quad \text{(Also see Ayeen Akbery, vol. 3, p. 32.)}$$

$$\frac{\text{D} - \sqrt{((\text{D} + c)(\text{D} - c))}}{2} = v.$$

$$2\sqrt{(\text{D} - v)} \times v = c.$$

$$\frac{4a\text{D}(c - a)}{\frac{5}{4}c^2 - (c - a)a} = c, \text{ and } \frac{c}{2} - \sqrt{\left(\frac{c^2}{4} - \frac{5}{4}c^2c}{4\text{D} + c}\right)} = a.$$

Also formulæ for the sides of the regular polygons of 3, 4, 5, 6, 7, 8, 9 sides inscribed in a circle. There are also rules for finding the area of a circle, and the surface and solidity of a sphere. It will be seen also that Bhascara is supposed to have given these two rules, viz—the sine of the sum of two arcs is equal to the sum of the products of the sine of each multiplied by the cosine of the other, and divided by the radius; and the cosine of the sum of two arcs is equal to the difference between the products of their sines and of their cosines divided by radius.

Is it to be doubted that the Hindoos applied their rule for the construction of the sines, to ascertain the ratio of the diameter of a circle to its circumference?—thus the circumference of a circle being divided into 360 degrees, or 21600 minutes, the sine of 90 degrees which is equal to the radius would be found by the rule 3438. This would give the ratio of the diameter to the circumference $7 : 21\frac{567}{573}$ and $1250 : 3926\frac{402}{573},$ and assuming, as the Hindoos commonly do, the nearest integers, the ratio would be 7 : 22 or 1250 : 3927.

It is not to be denied that there are some remarkable coincidences between the Greek and the Hindoo science ; for example, among many which might be given it may be suggested that the contrivances ascribed to Antiphon and Bryso, and that of Archimedes, for finding the ratio of the diameter of a circle to its circumference might have been the foundation of the Hindoo method; that Diophantus's speculations on indeterminate problems might be the origin of the Hindoo Algebra. But there are no truths in the history of science of which we are better assured than that the Surya Siddhanta rule for the sines, with the ratio of the diameter of a circle to its circumference 1250 : 3927; and the Bija Ganita rules for indeterminate problems were not known to the Greeks. Such are the stumbling blocks which we always find in our way when we attempt to refer the Hindoo science to any foreign origin.

Indians; and Massoudi refers Ptolemy's astronomy to them. (See Bailly's preface to his Indian astronomy, where is cited M. de Guignes Mem. Acad. Ins. T. 36, p. 771). Fyzee, who doubtless was conversant with Greco-arabian learning, and certainly knew the Hindoos well, has never started any doubt of the originality of what he found among them. The preface to the Zeej Mahommed Shahry, or Astronomical Tables, which were published in India in 1728, speaks of the European, the Greek, the Arabian, and the Indian systems as all different. That work was compiled with great learning by persons who were skilled in the sciences of the West, as well as those of the East *. More examples might be given—but to proceed.

The Bija Ganita is a Sanscrit treatise on algebra, by Bhascara Acharya, a celebrated Hindoo Astronomer and Mathematician.

Fyzee †, who, in 1587, translated the Lilavati, a work of his on arithmetic, mensuration, &c. speaks of an astronomical treatise of Bhascara s, dated in the 1105th year of the Salibahn, which answers to about 1183 of our æra; but Fyzee also says, it was 373 years before 995 Hegira, which would bring it down to A.D. 1225. So that Bhascara must have written about the end of the 12th century, or beginning of the 13th.

A complete translation of the Bija Ganita is a great desideratum; so it has been for more than 20 years, and so it seems likely to remain.

It will be seen however that we have already means of learning, with sufficient accuracy, the contents of this work. I have a Persian translation of the Bija Ganita, which was made in India in 1634, by Ata Allah Rusheedee. The Persian does not in itself afford a correct idea of its original, as a translation should do; for it is an

* See Asiatic Researches, 5th vol. on the Astronomical Labours of Jy Singh.

† I will here translate a part of Fyzee's preface:—"By order of king Akber, Fyzee translates into Persian, "from the Indian language, the book Lilavati, so famous for the rare and wonderful arts of calculation and "mensuration. He (Fyzee) begs leave to mention that the compiler of this book was Bhascara Acharya, whose "birth place, and the abode of his ancestors was the city of Biddur, in the country of the Deccan. Though "the date of compiling this work is not mentioned, yet it may be nearly known from the circumstance, that the "author made another book on the construction of Almanacks, called Kurrun Kuttohul, in which the date of "compiling it is mentioned to be 1105 years from the date of the Salibahn, an æra famous in India. From "that year to this, which is the 32d Ilahi year, corresponding with the lunar year 995, there have passed "373 years."

As the Ilahi began in the Hegira (or lunar) year, 992, (see Ayeen Akbery) the date 32 of Ilahi is of course an error. It is likely too that there is an error in the number 373.

Mr. Colebrooke, in the 9th vol. of Asiatic Researches, gives, on Bhascara's own authority, the date of his birth, viz. 1063 Saca. In 1105 Salibahn (or Saca) that is, about A.D. 1183, he was 42 years old.

undistinguished mixture of text and commentary, and in some places it even refers to Euclid. But to infer at once from this that every thing in the book was derived from Greek or Arabian writers, or that it was *inseparably* mixed, would be reasoning too hastily. A little patience will discover evidence of the algebra which it contains, being purely Hindoo*.

The following paper consists of an account of this translation, and some notes which I shall now mention:

Mr. Davis, the well-known author of two papers on Indian Astronomy in the Asiatic Researches, made, many years ago, in India, some abstracts and translations from the original Sanscrit Bija Ganita†, and it is greatly to be regretted that he did not complete a translation of the whole. The papers which contain his notes had long since been mislaid and forgotten. They have been but very lately found, and I gladly avail myself of Mr Davis's permission to make use of them here. The chief part of them is inserted at the end of my account of the Persian translation. To prevent misconception about these notes, it is proper for me to observe that in making them Mr. Davis had no other object than to inform himself generally of the nature of the Bija Ganita; they were not intended probably to be seen by any second person; certainly they were never proposed to convey a perfect idea of the work, or to be exhibited before the public in any shape. Many of them are on loose detached pieces of paper, and it is probable that, from the time they were written till they came into my hands, they were never looked at again. But nevertheless it will be seen that they do, without doubt, describe accurately a considerable portion of the most curious parts of the Bija Ganita; and though they may seem to occupy but a secondary place here, they will be found of more importance than all the rest of this work together.

They shew positively that the main part of the Persian translation is taken from

---

* The late Mr. Reuben Burrow in one of his papers in the Asiatic Researches says, he made translations of the Bija Ganita and Lilavati. Those translations he left to Mr. Dalby. They consist of fair copies in Persian of Ata Allah's and Fyzee's translations, with the English of each word written above the Persian. The words being thus translated separately, without any regard to the meaning of complete sentences, not a single passage can be made out. It is plain, from many short notes which Mr. Burrow has written in the margin of his Bija Ganita, that he made his verbal translation by the help of a Moonshee, and that he had the original Sanscrit at hand, with some opportunity of consulting it occasionally. I am much obliged to Mr. Dalby for allowing me the use of Mr. Burrow's copy which has enabled me to supply deficiencies in mine; and it is otherwise interesting, because it shews that Mr. Burrow had access to the original Sanscrit (probably by means of a Moonshee and a Pundit) and compared it with the Persian.

† It is to be remarked that they were made from the Sanscrit only. Mr. Davis never saw the Persian translation.

the Sanscrit work, and that the references to Euclid are interpolations of the Persian translator they give most of the Hindoo Algebraic notation* which is wanting in the Persian, and they shew that the Astronomy of the Hindoos was connected with their Algebra.

I must however confess, that even before I saw these notes the thing was to my mind quite conclusive. For I found (as will be seen) in this Persian translation of 1634, said to be from the Sanscrit, a perfectly connected structure of science, comprehending propositions, which in Europe were invented successively by Bachet de Mezeriac, Fermat, Euler, and De La Grange †.

---

* The Hindoos have no mark for +, they only separate the quantities to be added by a vertical line thus | or ||, as they separate their slocas or verses.

Their mark for *minus* is a dot over the quantity to be subtracted.

Instead of a mark for multiplication they write the factors together as we do, thus, ab for $a \times b$.

Division they mark as we do by a horizontal line drawn between the dividend and divisor, the lower quantity being the divisor.

For unknown quantities they use letters of the alphabet as we do. They use the first letters of the words signifying colours.

The known quantity (which is always a number) has the word roop (form) or the first letter of the word prefixed.

The square of the unknown quantity is marked by adding to the expression of the simple quantity the first letter of the word which means square, and in like manner the cube.

The sides of an equation are written one above another; every quantity on one side is expressed again directly under it on the other side. Where there is in fact no corresponding quantity, the form is preserved by writing that quantity with the co-efficient 0.

The methods of prefixing a letter to the known number, and using the first letter of the words square and cube are the same as those of Diophantus. I mention it as a curious coincidence; perhaps some people may attach more importance to it than I do.

'† The propositions which I here particularly allude to are these:—

1. A general method of solving the problem $\dfrac{ax \pm c}{b} = y$, a, b and c being given numbers and x and y indeterminate. The solution is founded on a division like that which is made for finding the greatest common measure of two numbers. The rules comprehend every sort of case, and are in all respects quite perfect.

2. The problem $am^2 + 1 = n^2$, (a being given and m and n required) with its solution.

3. The application of the above to find any number of values of $ax^2 + b = y^2$ from one known case.

4. To find values of x and y in $ax^2 + b = y^2$ by an application of the problem $\dfrac{ax \pm c}{b} = y$. It is unnecessary for me *here* to give any detail of the Hindoo methods.

The first question about this extraordinary matter is, what evidence have we that it is not all a forgery? I answer, shortly, that independently of its being now found in the Sanscrit books, it is ascertained to have been there in 1634 and 1587, that is to say, in times when it could not have been forged.

The following extract from a paper of De La Grange, in the 24th volume of the Memoirs of the Berlin Academy, for the year 1768, contains a summary of that part of the history of Algebra which is now alluded to. As for the 4th of the points abovementioned, the method in detail (however imperfect in some respects) is, as far as I know, new to this day. The first application of the principle in Europe is to be sought in the writings of De I a Grange himself.

To maintain that the Bija Ganita rules for the solution of indeterminate problems might have been had from any Greek or Arabian, or any modern European writers before the Mathematicians just named, would be as absurd as to say that the Newtonian Astronomy might have existed in the time of Ptolemy. It is true that Bachet wrote a few years before 1634, but this is no sort of objection to the argument, for that part which might be questioned as a mere copy of Bachet's method, namely, the rules for indeterminate problems of the first degree, is closely connected with matters of latter invention in Europe, and is in Mr. Dalby's copy of Fyzee's translation of the Lilavati, which I have before said was made in 1587; and Mr. Davis's notes shew that it is in the Sanscrit Bija Ganita, which was

---

" La plupart des Géometres qui ont cultivé l'analyse de Diophante se sont, a l'exemple de cet illustre inven-
" tuer, uniquement appliqués à eviter les valeurs irrationelles; et tout l'artifice de leurs méthodes se reduit à
" faire en sorte que les grandeurs inconnues puissent se déterminer par des nombres commensurables.

" L'art de resoudre ces sortes de questions ne demande gueres d'autres principes que ceux de l'analyse
" ordinaire : mais ces principes deviennent insuffisant lors-qu'on ajoute la condition que les quantités cher-
" chées soient non seulement commensurables mais encore égales à des nombres entiers.

" M. Bachet de Mezeriac, auteur d'un excellent commentarie sur Diophante et de différens autres ouvrages
" est, je crois, le premier quit ait tanté de soumettre cette condition au calcul. Ce savant a trouvé une méthode
" générale pour resoudre en nombres entiers toutes les equations du premier degré a deux ou plusieurs incon-
" nues, mais il ne paroit pas avoir été plus loin; et ceux qui aprés lui se sont occupés du même objet, ont
" aussi presque tous borné leurs recherches aux equations indéterminées du premier degré; leurs efforts se
" sont réduits a varier les méthods qui peuvent servir a la resolution de ces sortes d'equations, et aucun, si
" j'ose le dire, n'a donné une methode plus directe, plus générale, et plus ingenieuse que celle de M. Bachet
" qui se trouve dans ses récréations mathématiques intituéles ' *Problems plaisans et délectables qui se font par les*
" *nombres.*' Il est a la vérité assez surprenant que M. de Fermat qui s'etoit si long tems et avec tant de
" succés exercé sur la théorie des nombres entiers, n'ait pas cherché à resoudre généralement les problems
" indéterminés du second degré, et des degrés superieurs comme M. Bachet avoit fait ceux du premier degré;
" on a cependent lieu de croire qu'il s'etoit aussi appliqué a cette recherche, par le probleme qu'il proposa
" comme une espece de défi à M. Wallis et à tous les Geometres Anglois, et qui consistoit à trouver deux
" carrés entiers, dont l'un étant multiplié par un nombre entier donné non carré & ensuite retranché de l'autre,
" le reste fut etre égal à l'unité, car, outre que ce probleme est un cas particulier des équations du second
" dégré à deux inconnués il est comme la clef de la résolution génerale de ces équations. Mais soit que
" M. de Fermat n'ait pas continué ses recherches sur cette matiere, soit qu'elle ne soit parvenue jusqu'à nous,
" il est certain qu'on n'en trouve aucune trace dans ses ouvrages.

" Il paroit même que les Geometres Anglois qui ont résolu le probleme de M. de Fermat n'ont pas connu
" toute l'importance dont il est pour la solution générale des problemes indéterminés du second degré, du
" moins on ne voit pas qu'ils en ayant jamais fait usage, et Euler est si je ne me trompe, le prémier qui ait
" fait voir comment à l'aide de ce probleme on peut trouver une infinité de solutions en nombres entiers de
" toute équation du second dégré à deux inconnues, dont on connoit déja une solution.

" Il résulte de tout ce que nous venons de dire, que depuis l'ouvrage de M. Bachet que a paru en 1613,
" jusqu'a présent, ou du moins jusqu'au mémoire que je donnai l'année passée sur la solution de problems
" indéterminés du second dégré, la théorie de ces sortes de problemes n'avoit pas a proprement parler, été
" poussée au dela du premier dégré."

written four centuries before Bachet. Though we are not without direct proofs from the original, yet, as even the best Sanscrit copies of the Bija Ganita, or any number of such copies exactly corresponding, would still be open to the objection of interpolations, it is necessary in endeavouring to distinguish the possible and the probable corruptions of the text, from what is of Indian origin, to recur to the nature of the propositions themselves, and to the general history of the science. Indeed we have not data enough to reason satisfactorily on other principles. We cannot rely upon the perfect identity or genuineness of any book before the invention of printing, unless the manuscript copies are numerous, and of the same age as the original. Such is the nature of our doubt and difficulty in this case, for old mathematical Sanscrit manuscripts are exceedingly scarce; and our uncertainty is greatly increased by a consideration of this fact, that in latter times the Greek, Arabian, and modern European science has been introduced into the Sanscrit books.

Yet, in cases precisely parallel to this of the Hindoos, we are not accustomed to withhold our belief as to the authenticity of the reputed works of the ancients, and in forming our judgment we advert more to the contents of the book than to the state of the manuscript. When the modern Europeans first had Euclid, they saw it only through an Arabic translation. Why did they believe that pretended translation to be authentic? Because they found it contained a well connected body of science; and it would have been equally as improbable to suppose that the Arabian translator could have invented it himself as that he could have borrowed it from his countrymen. There are principles on which we decide such points. We must not look for mathematical proof, but that sort of probability which determines us in ordinary matters of history.

Every scrap of Hindoo science is interesting; but it may be asked why publish any which cannot be authenticated? I answer, that though this translation of Ata Allah's which professes to exhibit the Hindoo algebra in a Persian dress, does indeed contain some things which are not Hindoo, yet it has others which are certainly Hindoo. By separating the science from the book we may arrive at principles, which if cautiously applied, cannot mislead, which in some cases will shew us the truth, and will often bring us to the probability when certainty is not to be had. On this account I think the Persian translation at large interesting, notwithstanding it contains some trifling matters, some which are not intelligible, and others which are downright nonsense.

I have said that Mr. Davis's notes shew a connexion of the algebra of the

Hindoos with their astronomy. Mr. Davis informs me that in the astronomical treatises of the Hindoos, reference is often made to the algebra; and particularly he remembers a passage where Bhascara says " it would be as absurd for a person " ignorant of algebra to write about astronomy, as for one ignorant of grammar " to write poetry."

Bhascara, who is the only Hindoo writer on algebra whose works we have yet procured, does not himself pretend to be the inventor, he assumes no character but that of a compiler*. Fyzee never speaks of him but as a person eminently skilled in the sciences he taught. He expressly calls him the compiler of the Lilavati.

I understand from Mr. Davis, and I have heard the same in India, that the Bija Ganita was not intended by Bhascara as a separate unconnected work, but as a component part of one of his treatises on astronomy, another part of which is on the circles of the sphere.

I have found among Mr. Davis's papers, some extracts from a Sanscrit book of astronomy, which I think curious, although the treatise they were taken from is modern. Mr. Davis believes it to have been written in Jy Sing's time, when the European improvements were introduced into the Hindoo books. Two of these extracts I have added to the notes on the Bija Ganita. The first of the two shews that a method has been ascribed by Hindoo Astronomers to Bhascara of calculating sines and cosines by an application of the principles which solve indeterminate problems of the second degree. This suggestion is doubtless of Hindoo origin, for the principles alluded to were hardly known in Europe in Jy Sing's time†. I think it very probable that the second extract is also purely Hindoo, and that the writer knew of Hindoo authors who said the square root might be extracted by the cootuk; that is to say, the principle which effects the solution of indeterminate problems of the first degree. From this, and from what is in the Bija Ganita, one cannot but suspect that the Hindoos had continued fractions, and possibly some curious arithmetic of sines. On such matters however, let every one exercise his own judgment. ‡

---

* " Almost any trouble and expence would be compensated by the possession of the three copious treatises " on algebra from which Bhascara declares he extracted his Bija Ganita, and which in this part of India are " supposed to be entirely lost."—As. Res. vol. iii. Mr. Davis " On the Indian Cycle of 60 years."

† Jy Sing reigned from 1694 to 1744.

‡ Mr. Reuben Burrow, who, by the bye, it must be confessed is very enthusiastic on these subjects, in a paper

We must not be too fastidious in our belief, because we have not found the works of the teachers of Pythagoras; we have access to the wreck only of their ancient learning; but when we see such traces of a more perfect state of knowledge, when we see that the Hindoo algebra 600 years ago had in the most interesting parts some of the most curious modern European discoveries, and when we see that it was at that time applied to astronomy, we cannot reasonably doubt the originality and the antiquity of mathematical learning among the Hindoos. Science in remote times we expect to find within very narrow limits indeed. its *history* is all we look to in such researches as these. Considering this, and comparing the contents of the Hindoo books with what they might have been expected to contain, the result affords matter of the most curious speculation.

May I be excused for adding a few words about myself. If my researches have not been so deep as might have been expected from the opportunities I had in India, let it be remembered that our labours are limited by circumstances. It is true I had at one time a copy of the original Bija Ganita, but I do not understand Sanscrit, nor had I then any means of getting it explained to me. Official avocations often prevented me from bestowing attention on these matters, and from seizing opportunities when they did occur. Besides, what is to be expected in this way from a *mere amateur*, to whom the simplest and most obvious parts only of such subjects are accessible?

E. S.

The following account of Ata Allah's Bija Ganita is partly literal translation, partly abstract, and partly my own.

The literal translation is marked by inverted commas; that part which consists of my own remarks or description will appear by the context, and all the rest is abstract.

I have translated almost all the rules, some of the examples entirely, and

---

in the appendix of the 2d vol. of the As. Res. speaks of the Lilavati and Bija Ganita, and of the mathematical knowledge of the Hindoos: He says, he was told by a Pundit, that some time ago there were other treatises of algebra, &c. (See the paper.)

others in part; in short, whatever I thought deserving of particular attention, for the sake of giving a distinct idea of the book.

Perhaps some of the translated parts might as well have been put in an abstract; the truth is, that having made them originally in their present form I have not thought it worth while to alter them.

The notes are only a few remarks which I thought might be of use to save trouble and to furnish necessary explanation.

BIJA GANITA.

"After the usual invocations and compliments, the Persian translator begins thus:
"By the Grace of God, in the year 1044 Hegira" (or A. D. 1634) "being the
"eighth year of the king's reign, I, Ata Alla Rasheedee, son of Ahmed Nadir,
"have translated into the Persian language, from Indian, the book of Indian
"Algebra, called Beej Gunnit (Bija Ganita), which was written by Bhasker Acharij
"(Bhascara Acharya) the author of the Leelawuttee (Lilavati). In the science of
"calculation it is a discoverer of wonderful truths and nice subtilties, and it con-
"tains useful and important problems which are not mentioned in the Leelawuttee,
"nor in any Arabic or Persian book. I have dedicated the work to Shah Jehan, and
"I have arranged it according to the original in an introduction and five books."

INTRODUCTION.

"The introduction contains six chapters, each of which has several sections."

CHAPTER I.
On Possession (مال)* and Debt (دين).

"Know that whatever is treated of in the science of calculation is either
"affirmative or negative; let that which is affirmative be called *mal*, and that
"which is negative *dein*. This chapter has five sections."

Sect. I.

On Addition and Subtraction, that is, to encrease and diminish.

"If an affirmative is taken from an affirmative, or a negative from a negative,
"the subtrahend is made contrary; that is to say, if it is affirmative suppose it
"negative, and if negative suppose it affirmative, and proceed as in addition.
"The rule of addition is, that if it is required to add two affirmative quantities,

---

* Most of the technical terms here used are Arabic.

" or two negative quantities together, the sum is the result of the addition. If
" they are affirmative call the sum affirmative: if negative call the sum negative.
" If the quantities are of different kinds take the excess; if the affirmative is
" greater, the remainder is affirmative; if the negative is greater, the remainder
" is negative; and so it is in subtraction." (Here follow examples).

SECT. II.

On Multiplication *.

" If affirmative is multiplied by affirmative, or negative by negative, the product
" is affirmative and to be included in the product. If the factors are contrary
" the product is negative, and to be taken from the product. For example, let us
" multiply two affirmative by three affirmative, or two negative by three negative,
" the result will be six affirmative; and if we multiply two affirmative by three
" negative, or the contrary, the result will be six negative."

SECT. III.

On Division.

" The illustration of this is the same as what has been treated of under multi-
" plication, that is to say, if the dividend and the divisor are of the same kind
" the quotient will be affirmative, and if they are different, negative. For
" example, if 8 is the dividend and 4 the divisor, and both are of the same kind,
" the quotient will be 2 affirmative; if they are different, 2 negative."

SECT. IV.

On Squares†.

" The squares of affirmative and negative are both affirmative; for to find the

---

* In the Persian translation the product of numbers is generally called the rectangle.

† I had a Persian treatise on Algebra in which there was this passage—" Any number which is to be multiplied
" by itself is called by arithmeticians root (جذر), by measurers of surfaces side (ضلع), and by alge-
" braists thing (شي). And the product is called by arithmeticians square (مجذور), by measurers
" of surfaces square (مربع), and by algebraists possession (مال)." مال is also used for *plus*, and
its opposite debt (دين) for *minus*. These terms, all of which are Arabic, are used in the Persian translation
of the Bija Ganita, the geometrical more frequently than their corresponding arithmetical or algebraical ones.

" square of 4 affirmative we multiply 4 affirmative by 4 affirmative, and by the rules
" of multiplication, as the factors are of the same kind, the product must be 16
" affirmative, and the same applies to negative."

Sect. V.

On the Square Root.

" The square root of affirmative is sometimes affirmative and sometimes nega-
" tive, according to difference of circumstances. The square of 3 affirmative or
" of 3 negative is 9 affirmative; hence the root of 9 affirmative is sometimes 3
" affirmative and sometimes 3 negative, according as the process may require.
" But if any one asks the root of 9 negative I say the question is absurd, for there
" never can be a negative square as has been shown."

CHAP. II.

On the Cipher.

" It is divided into four sections."

Sect. I.

On Addition and Subtraction.

" If cipher is added to a number, or a number is added to cipher, or if cipher
" is subtracted from a number, the result is that number: and if a number is sub-
" tracted from cipher, if it is affirmative it becomes negative, and if negative it
" becomes affirmative. For example, if 3 affirmative is subtracted from cipher
" it will be 3 negative, and if 3 negative is subtracted it will be 3 affirmative."

Sect. II.

On Multiplication.

" If cipher is multiplied by a number, or number by cipher, or cipher by cipher,
" the result will be cipher. For example, if we multiply 3 by cipher, or con-
" versely, the result will be cipher"

Sect. III.

On Division.

" If the dividend is cipher and the divisor a number the quotient will be cipher.
" For example, if we divide cipher by 3 the quotient will be cipher, for multi-
" plying it by the divisor the product will be the dividend, which is cipher:
" and if a number is the dividend and cipher the divisor the division is impossible;
" for by whatever number we multiply the divisor, it will not arrive at the divi-
" dend, because it will always be cipher."

Sect. IV.

On Squares, &c.

" The square, cube, square root, and cube root of cipher, are all cipher; the
" reason of which is plain."

CHAP. III.

On Colours.

" Whatever is unknown in examples of calculation, if it is one, call it thing,
" (شي), and unknown (مجهول); and if it is more call the second black,
" and the third blue, and the fourth yellow, and fifth red. Let these be termed
" colours, each according to its proper colour.   This chapter has five sections.

Sect. I.

On Addition and Subtraction of Colours.

" When we would add one to another, if they are of the same kind add the
" numbers* together; if they are of two or more kinds, unite them as they are,
" and that will be the result of the addition."   Here follows an example.
" If we wish to subtract, that is to take one from the other, let the subtrahend
" be reversed.   If then two terms of the same kind are alike in this, that they are
" both affirmative or both negative, let their sum be taken, otherwise their dif-
" ference, and whatever of the kind cannot be got from the minuend, must be

---

* Meaning here the co-efficients.

" subtracted from cipher. Then let it be reversed, and this will be the result
" exactly." (Here follows an example).

SECT. II.

On Multiplication of Colours.

" If a colour is multiplied by a number the product will be a number*, $x \times x$
" will be x^2, whether the number is the same or different, and the product multi-
" plied by x will be x^3. If the colours are different multiply the numbers of both
" together, and call the product the rectangle of those two colours." The
following is given as a convenient method of multiplying:

	$+ 3x$	$+ 2$
$+5x$	$+ 15x^2$	$+ 10x$
-1	$- 3x$	$- 2$
Product	$+ 15x^2 + 7x - 2$	

which shews the product of $(5x - 1) \times (3x + 2.)$ (Here follow examples).

SECT. III.

On Division of Colours.

" Write the dividend and divisor in one place, find numbers or colours or both,
" such that when they are multiplied by the divisor, the product subtracted from
" the dividend will leave no remainder. Those numbers or colours will be the
" quotient."

---

* In the Persian translation there is no algebraic notation, I mean to] translate " the unknown" by x, " the
black" by y, and so on. And in like manner I have used the marks of multiplication, &c. instead of writing
the words at length as they are in the Persian.

SECT. IV.

On the square of Colours:

" That is to say, the product arising from any thing multiplied by itself."
Examples.

SECT. V.

On the Square Root of Colours.

" To know the square root of a colour, find that which when it is multiplied by
" itself the product subtracted from the colour whose root is required, will leave no
" remainder.   The rule is the same if there are other colours or numbers with
" that colour."

Example.   Required the square root of $16x^2+36-48x$.   The roots of $16x^2$ and
36 are $4x$ and 6, and as $48x$ is — these two roots must have different signs.
Suppose one + and the other —, multiply them and the product will be $-24x$;
twice this is $-48x$ which was required.   The root then is $+4x-6$, or $+6-4x$.

Another Example.   Required the square root of $9x^2+4y^2+z^2+12xy-6xz-$
$4yz-6x-4y+2z+1$.   Take the root of each square ; we have $3x$, $2y$, z, and 1.
Multiply these quantities and dispose the products in the cells of a square.

	$3x$	$2y$	z	1
$3x$	$9x^2$	$6xy$	$3xz$	$3x$
$2y$	$6xy$	$4y^2$	$2yz$	$2y$
z	$3xz$	$2yz$	z^2	z
1	$3x$	$2y$	z	1

To find what sort of quantities these are: The product of x and y is +, therefore the factors are like, suppose them both —. The product of x and z is —, therefore the former having been supposed — the latter must be + because the factors must be different. $3x$ is the product of $3x$ and 1; and x being —, 1 must be +. The sorts thus found are to be placed in the cells accordingly. The sum of the products is the square whose root was required. If x had been supposed + the sorts would have been contrary, the reason of which is plain.

CHAP. IV.

ON SURDS.

Containing five sections.

SECT. I.

On Addition and Subtraction.

To find the sum or difference of two surds; $\sqrt{a}$ and $\sqrt{b}$ for instance.

Rule. Call $a + b$ the greater surd; and if $a \times b$ is rational call $2\sqrt{ab}$ the less surd. The sum will be $\sqrt{(a+b+2\sqrt{ab})}$*, and the difference $\sqrt{(a+b-2\sqrt{ab})}$. If $a \times b$ is irrational the addition and subtraction are impossible.

Example. Required the sum of $\sqrt{2}$ and $\sqrt{8}$; $2 + 8 = 10$ the greater surd. $2 \times 8 = 16$, $\sqrt{16} = 4$, $4 \times 2 = 8$ the less surd. $10 + 8 = 18$ and $10 - 8 = 2$. $\sqrt{18}$ then will be the sum and $\sqrt{2}$ the difference. If one of the numbers is rational take its square and proceed according to the rule, and this must be attended to in multiplication and division, for on a number square with a number not square the operation cannot be performed.

Another Rule. Divide a by b and write $\sqrt{\dfrac{a}{b}}$ in two places In the first place add 1, and in the second subtract 1; then we shall have $\sqrt{\left(\left(\sqrt{\dfrac{a}{b}}+1\right)^2 \times b\right)}$ $= \sqrt{a} + \sqrt{b}$ and $\sqrt{\left(\left(\sqrt{\dfrac{a}{b}}-1\right)^2 \times b\right)} = \sqrt{a} - \sqrt{b}$. If $\dfrac{a}{b}$ is irrational the addition can only be made by writing the surds as they are, and the subtraction by writing the greater number + and the less —.

---

* For $\sqrt{(a + b \pm 2\sqrt{ab})} = \sqrt{a} \pm \sqrt{b}$.

Sect. II.

On Multiplication.

Proceed according to the rules already given; but if one of the factors has numbers as dirhems or dinars, take their squares and go on with the operation.

Example. Multiply $\sqrt{3}+5$ by $\sqrt{2}+\sqrt{3}+\sqrt{8}$ As 5 is of the square sort take its square, and arrange in a table thus :

	$\sqrt{2}$	$\sqrt{3}$	$\sqrt{8}$
$\sqrt{3}$	$\sqrt{6}$	$\sqrt{9}$	$\sqrt{24}$
$\sqrt{25}$	$\sqrt{50}$	$\sqrt{75}$	$\sqrt{200}$
Product	$3+\sqrt{54}+\sqrt{450}+\sqrt{75}$		

In summing the terms of the product, if any square number is found, take its root. Here 9 is found and its root is 3. The rest of the terms being irrational, add such as can be added. $\sqrt{6}+\sqrt{24}=\sqrt{54}$. If this last were a square number its root should be extracted.

Again, $\sqrt{50}+\sqrt{200}=\sqrt{450}$. No further addition is possible; the complete product therefore is $3+\sqrt{54}+\sqrt{450}+\sqrt{75}$.

Another rule to be observed is, if any of the terms which compose the factors can be added, take their sum and write it in the table instead of the terms of which it is formed. Thus in the last example $\sqrt{2}$ and $\sqrt{8}$ may be added. Write $\sqrt{18}$ which is their sum in the table, and we shall have

	$\sqrt{3}$	$\sqrt{18}$
$\sqrt{3}$	$\sqrt{9}$	$\sqrt{54}$
$\sqrt{25}$	$\sqrt{75}$	$\sqrt{450}$
	$3+\sqrt{75}+\sqrt{54}+\sqrt{450}$	

and the result is the same as before.

Another Example. Multiply $\sqrt{3}+\sqrt{25}$ by $\sqrt{3}+\sqrt{12}-5$. Instead of $\sqrt{3}$ and $\sqrt{12}$ write their sum $\sqrt{27}$. Take the square of 5 it is 25, and this is negative notwithstanding the rule which says that whether the root is negative or affirmative the square shall be affirmative. Here the square must be of the same sort as the root. Multiply $\sqrt{27}-\sqrt{25}$ by $\sqrt{3}+\sqrt{25}$.

	$+\sqrt{27}$	$-\sqrt{25}$
$+\sqrt{3}$	$+\sqrt{81}$	$-\sqrt{75}$
$+\sqrt{25}$	$+\sqrt{675}$	$-\sqrt{625}$
	$-16+\sqrt{300}$	

$\sqrt{81}=9$ and $\sqrt{625}=25$. 25 being negative and 9 affirmative their sum is -16, and the sum of $+\sqrt{675}$ and $-\sqrt{75}$ is $+\sqrt{300}$. Therefore $(\sqrt{3}+\sqrt{25})\times(\sqrt{3}+\sqrt{12}-5)=-16+\sqrt{300}$.

SECT. III.

On Division.

Divide the dividend by the divisor, and if the quotient is found without a remainder the division is complete. When this cannot be done proceed as follows:

When in the divisor there are both affirmative and negative terms, if there are more of the former make one of them negative; if more of the latter make one of them affirmative. When all the terms are affirmative make one negative, and when all are negative make one affirmative. When the number of affirmative terms is equal to that of the negative, it is optional to change one of them or not. Multiply the divisor (thus prepared) by the original divisor, and add the products rejecting such quantities as destroy each other. Multiply the prepared divisor by the dividend, and divide the product of this multiplication by that of the former the result will be the quotient required.

Example. Let the dividend be that which was the product in the first example under the rule for multiplication, viz. $3 + \sqrt{54} + \sqrt{450} + \sqrt{75}$, and the divisor $\sqrt{18} + \sqrt{3}$.

$$\frac{75}{3} = 25, \quad \frac{450}{18} = 25, \quad 3^2 = 9, \quad \frac{9}{3} = 3, \quad \frac{54}{18} = 3, \quad \sqrt{25} = 5,$$

the quotient then is $5 + \sqrt{3}$.

Another Example. Divide $\sqrt{9} + \sqrt{54} + \sqrt{450} + \sqrt{75}$ by $5 + \sqrt{3}$. Make $\sqrt{3}$ negative, and multiply 5 (or $\sqrt{25}$) $- \sqrt{3}$ by the divisor $\sqrt{25} + \sqrt{3}$.

	$+\sqrt{25}$	$-\sqrt{3}$
$+\sqrt{3}$	$+\sqrt{75}$	$-\sqrt{9}$
$+\sqrt{25}$	$+\sqrt{625}$	$-\sqrt{75}$

$\sqrt{75}$ occurring twice with opposite signs is destroyed. $\sqrt{625}=25$, $\sqrt{9}=-3$, $25-3=22=\sqrt{484}$. Multiply $\sqrt{25}-\sqrt{3}$ by the dividend and we have

	$+\sqrt{9}$	$+\sqrt{54}$	$+\sqrt{450}$	$+\sqrt{75}$
$-\sqrt{3}$	$-\sqrt{27}$	$-\sqrt{162}$	$-\sqrt{1350}$	$-\sqrt{225}$
$+\sqrt{25}$	$+\sqrt{225}$	$+\sqrt{1350}$	$+\sqrt{11250}$	$+\sqrt{1875}$

Here $\sqrt{225}$ and $\sqrt{1350}$ are rejected. Find the sum of $\sqrt{27}$ and $\sqrt{1875}$ in this manner,

$1875 + 27 = 1902$, $1875 \times 27 = 50625$, $\sqrt{50625} = 225$

$225 \times 2 = 450$, $1902 - 450 = 1452$, $\sqrt{1452} = \sqrt{1875} - \sqrt{27}$.

Next find the sum of $\sqrt{162}$ and $\sqrt{11250}$.

$162 + 11250 = 11412$, $162 \times 11250 = 1822500$,

$\sqrt{1822500} = 1350$, $1350 \times 2 = 2700$, $11412 - 2700 = 8712$,

$\sqrt{8712} = \sqrt{11250} - \sqrt{162}$. By the multiplication of the dividend we have found $\sqrt{1452}$ and $\sqrt{8712}$.

Divide these by $\sqrt{484}$ which was the result of the multiplication of the divisor, and we shall have $\sqrt{18}$, and $\sqrt{3}$ for the quotient required. If $\sqrt{3}$ is retained as correct, and $\sqrt{18}$ is considered as incorrect, instead of $\sqrt{18}$ other numbers may be found by the following rule *.

Divide the incorrect number (meaning the number under the radical sign) by any square number which will divide it without a remainder, and note the quotient. Divide the root of that square number into as many parts as there are numbers required. Take the squares of these parts; multiply them by the quotient above

---

* To resolve $\sqrt{a}$ into several parts, divide a by any square b^2, and let b be resolved into as many parts, c, d, e, &c. as may be required. Then $\sqrt{a} = \sqrt{\left(\frac{a}{b^2}c^2\right)} + \sqrt{\left(\frac{a}{b^2}d^2\right)} + \sqrt{\left(\frac{a}{b^2}e^2\right)}$ &c. which may be proved by adding the quantities.

found, and the roots of the several products will be the remaining parts of the quotient required.

$$\frac{18}{9} = 2, \quad \sqrt{9} = 3, \quad 3 = 1 + 2, \quad 1^2 = 1, \quad 2^2 = 4, \quad 1 \times 2 = 2, \quad 4 \times 2 = 8.$$

$\sqrt{2}$ and $\sqrt{8}$ are the remaining parts of the quotient.

SECT IV.

On the Squares of Surds.

Multiply the surds by themselves.—(Here follow examples).—The squares are found by multiplying the surds in the common way.

SECT. V.

On finding the Square Roots of the Squares of Surds *.

" If the square is of one surd or more, and I would find its root; first I take the
" square of the numbers that are with it, and subtract these squares from it.
" Accordingly after subtraction something may remain.   I take the root of what-
" ever remains, add it in one place to the original number, and in another sub-
" tract it from the same.   Halve both the results, and two roots will be obtained.
" I then re-examine the squares of the surds to know whether any square remains

---

* Let $a + \sqrt{b} + \sqrt{c} + \sqrt{d}$, &c. be the square of a multinomial surd, a the sum of the squares of the roots, and $\sqrt{b} + \sqrt{c} + \sqrt{d} + $ &c. the product of the roots taken two and two. The number of roots being n, the number of terms in the square will be n^2, of which n will be the number of rational terms, and $n^2 - n$ the number of surd products.   If we call the double products single terms, $\dfrac{n^2 - n}{2}$ will express the number of surd terms, and considering the sum of the rational terms as one term, the proposed square may be reduced to the form

$$(x + y + z + \&c.) + (2\sqrt{xy} + 2\sqrt{xz} + \&c.) + (2\sqrt{yz} + \&c. \ \&c.)$$

where $\sqrt{x} + \sqrt{y} + \sqrt{z} + $ &c. is the root of the square, and the surd terms of the square are divided into periods of $n - 1$, $n - 2$, $n - 3$, &c. as directed in the Beej Gunnit.

$$\text{Supposing } x + y + z + \&c. = Q$$
$$y + z + \&c. = R$$
$$z + \&c. = S$$
$$\&c. \quad \&c.$$

$$\sqrt{\frac{Q \pm \sqrt{(Q^2 - 4xR)}}{2}} = \sqrt{x} \text{ or } \sqrt{R}$$

$$\sqrt{\frac{R \pm \sqrt{(R^2 - 4yS)}}{2}} = \sqrt{y} \text{ or } \sqrt{S}$$

$$\sqrt{\frac{S \pm \sqrt{(S^2 - 4zT)}}{2}} = \sqrt{z} \text{ or } \&c. \text{ and so on.}$$

" after the subtraction or not: if none remains these two are the roots required;
" if any remains, that one of these two roots, to which the following rule cannot
" be applied, is correct, and the other is the sum of two roots; from that root we
" obtain the two roots required. The way of the operation is this, suppose that
" root number, and take its square, and subtract from it the square which was
" not subtracted at first, and take the root of the remainder; let this be added in
" one place to the original number which we supposed, and subtracted from it in
" another place, and halve both the results, two roots will be obtained. If then
" these three are the roots required, the operation is ended, otherwise go on with
" it in the same manner till all the roots are found; and if the first question is
" of a number without a square of a surd, it may be solved by the operation
" which was described at the end of division. And if in the square there are one
" or more surds negative, suppose them affirmative, and proceed to the end with
" the operation; and of the two roots found let one be negative."

Required the root of $5 + \sqrt{24}$; $5^2 = 25$, $25 - 24 = 1$, $\sqrt{1} = 1$, $5 + 1 = 6$;
$5 - 1 = 4$, $\frac{6}{2} = 3$ and $\frac{4}{2} = 2$; and $\sqrt{3} + \sqrt{2} = \sqrt{(5 + \sqrt{24})}$.

Another Example. Required the root of $10 + \sqrt{24} + \sqrt{40} + \sqrt{60}$; $10^2 = 100$,
$100 - (24 + 40) = 36$, $\sqrt{36} = 6$, $10 + 6 = 16$, $10 - 6 = 4$, $\frac{16}{2} = 8$, $\frac{4}{2} = 2$, then
we have $\sqrt{8}$ and $\sqrt{2}$. As 60 remains to be subtracted, one of these two numbers
is one term of the root, and the other is the sum of two remaining terms
(should be the root of the sum of the squares of the remaining terms). The rule
is not applicable to 2, therefore 8 must be the sum of the terms. $8^2 = 64$,
$64 - 60 = 4$, $\sqrt{4} = 2$, $8 + 2 = 10$, $8 - 2 = 6$, $\frac{10}{2} = 5$ and $\frac{6}{2} = 3$. Wherefore $\sqrt{2} + \sqrt{3} + \sqrt{5} = \sqrt{(10 + \sqrt{24} + \sqrt{40} + \sqrt{60})}$.

Another Example. Required the root of $16 + \sqrt{24} + \sqrt{40} + \sqrt{48} + \sqrt{60} + \sqrt{72} + \sqrt{120}$; $16^2 = 256$, $256 - (24 + 40 + 48) = 144$, $\sqrt{144} = 12$, $16 + 12 = 28$,
$16 - 12 = 4$, $\frac{28}{2} = 14$, $\frac{4}{2} = 2$; we have then $\sqrt{14}$ and $\sqrt{2}$. As the rule does not
apply to 2, 14 must be the sum of two remaining terms of the root. $14^2 = 196$,
$196 - (120 + 72) = 4$, $\sqrt{4} = 2$, $14 + 2 = 16$, $14 - 2 = 12$, $\frac{16}{2} = 8$, $\frac{12}{2} = 6$. One surd
remaining, and the rule not being applicable to 6, 8 must be the sum of two

D

terms. $8^2=64$, $64-60=4$, $\sqrt{4}=2$, $8+2=10$, $8-2=6$, $\frac{10}{2}=5$, and $\frac{6}{2}=3$.
All the terms of the square having been brought down, the complete root is $\sqrt{6}+\sqrt{5}+\sqrt{3}+\sqrt{2}$.

Another Example. Required the root of 72: $72^2=5184$, $0^2=0$, $5184-0=5184$, $\sqrt{5184}=72$, $72+72=144$, $72-72=0$, $\frac{144}{2}=72$, $\frac{0}{2}=0$, $\sqrt{72}$ then is the root.

If instead of one term three terms are required, find them by the rule given in the section on division; divide by 36 which is a square number, $\frac{72}{36}=2$, $\sqrt{36}=6$, $6=3+2+1$, $3^2=9$, $2^2=4$, $1^2=1$, $9\times\frac{72}{36}=18$, $4\times\frac{72}{36}=8$, $1\times\frac{72}{36}=2$; therefore $\sqrt{72}=\sqrt{18}+\sqrt{8}+\sqrt{2}$. If three equal terms had been required, the root of the divisor must have been divided into three equal parts.

Another Example. It is required to find the difference of $\sqrt{3}$ and $\sqrt{7}$. The rule not being applicable to this case, suppose $\sqrt{7}$ affirmative, and $\sqrt{3}$ negative, the square of these numbers is $10-\sqrt{84}$. To determine the root of this, suppose 84 to be positive; $10^2=100$, $100-84=16$, $\sqrt{16}=4$, $10+4=14$, $10-4=6$, $\frac{14}{2}=7$, $\frac{6}{2}=3$. We have then $\sqrt{7}$ and $\sqrt{3}$, one of which must be *minus* because $\sqrt{84}$ was *minus*.

Another Example. Whether the root is $+\sqrt{2}+\sqrt{3}-\sqrt{5}$ or $-\sqrt{2}-\sqrt{3}+\sqrt{5}$ the square will be the same, viz. $10+\sqrt{24}-\sqrt{40}-\sqrt{60}$.

Let the root of this square be determined. $10^2=100$, $100-(40+60)=0$, $\sqrt{0}=0$, $10+0=10$, $10-0=10$, $\frac{10}{2}=5$, $\frac{10}{2}=5$. As $\sqrt{24}$ remains, $5^2=25$, $25-24=1$, $\sqrt{1}=1$, $5+1=6$, $5-1=4$, $\frac{6}{2}=3$, $\frac{4}{2}=2$. If $24+40$ is subtracted from 100 there remains 36, $\sqrt{36}=6$, $10+6=16$, $10-6=4$, $\frac{16}{2}=8$, $\frac{4}{2}=2$. As $\sqrt{60}$ remains $8^2=64$, $64-60=4$, $\sqrt{4}=2$, $8+2=10$, $8-2=6$, $\frac{10}{2}=5$, $\frac{6}{2}=3$. If $24+60$ is subtracted from 100 there remains 16. $\sqrt{16}=4$, $10+4=14$, $10-4=6$, $\frac{14}{2}=7$, $\frac{6}{2}=3$, $\sqrt{40}$ yet remaining, $7^2=49$, $49-40=9$, $\sqrt{9}=3$, $7+3=10$, $7-3=4$, $\frac{10}{2}=5$, $\frac{4}{2}=2$. The terms of the root are $\sqrt{2}$ and

$\sqrt{}$3 and $\sqrt{}$5. If 2 and 5, or 3 and 5 are both negative or both affirmative the operation will be the same; the only difference will be in the signs.

"If the root consists of one term* only, its square will be of the kind of
"number; if of two terms, its square will be number and one surd; if the root
"has three terms, the square will have one number and three surds; if it has four,
"the square will have one number and six surds; if five, one number and ten
"surds; and if six, one number and fifteen surds. The rule is, add the numbers
"in the natural scale, from 1 to the number next below that which expresses
"the number of terms in the root, the sum will shew the number of surds. For
"the use of beginners is annexed a table in which the first column shews the
"number of terms of the roots; the second column shews the number of surd
"terms in the squares; and the third the number of rational terms in the
"squares, from 1 to 9.

No. of terms in the root.	No. of surds in the square.	No. of rational terms in the squares.
1	0	1
2	1	1
3	3	1
4	6	1
5	10	1
6	15	1
7	21	1
8	28	1
9	36	1

* The number of surd terms in the square being $\frac{n^2-2}{2}$, is = the sum of the numbers in the natural scale from 1 to the number next below n.

" For numbers consisting of more terms than 9 the number of surds in the
" squares may be found by the rule which has been given.   If in the square there
" are three surd terms, first subtract two of them from the square of the numbers
" and afterwards subtract the third.   If there are six surds, first subtract 3, then
" 2, and so on ; if there are 10 surds, first subtract 4; if 15, first 5 ; if 21, first 6 ;
" if 28, first 7; if 36, first 8 ; and in general the number of surds of the square
" will be found in the table in the column of roots next above the number
" of its root.   If they are not subtracted in the regular order, the result will be
" wrong.   The test of the operation of this : if either of the two numbers found
" by the rule is multiplied by 4, and the number which was subtracted from the
" square of the rational term is divided by the product, the quotient will be the
" other number found, without any remainder.   If either of those two numbers
" is a correct term of the root, and the other the sum of two roots, the least, or
" that which is the correct term, whether in number it be more or less than the
" number of the sum of two roots, must be multiplied by 4, and every quantity
" that has been subtracted must be divided by the products, the quotient will be
" the numbers of the required roots from the second number.   If, after this divi-
" sion, there is any remainder the operation is wrong.

" The squares of all moofrid numbers* are made up either of rational numbers
" alone, or of rational numbers and surds, as has been seen in the examples of the
" section on squares.

" If a surd occurs there must be a moofrid number with it, otherwise its root
" cannot be found.   If a surd is divided into two :—For example, if $\sqrt{18}$ is
" divided into $\sqrt{2}$ and $\sqrt{8}$, its root will have one term more than it would have
" had regularly ; and if two surds are united the root will have one term less.
" These two operations of separation and union must be attended to and applied
" whenever they are possible."

Example.   Required the root of $10 + \sqrt{32} + \sqrt{24} + \sqrt{8}$.   From the square
of 10 which is 100, subtract any two of the numbers under the radical signs, and
the remainder will be irrational: the case is, therefore impossible.   If we proceed
contrary to the rule, by subtracting at once the three terms from 100, we shall
have 36 the remainder, then $\sqrt{36}=6$, $10+6=16$, $10-6=4$, $\frac{16}{2} = 8$, $\frac{4}{2} = 2$.   We

---

* Moofrid means simple as opposed to compound, but in the language of this science it is generally used to
express a number having one significant figure.

find then $\sqrt{8}$ and $\sqrt{2}$, but these are not the true roots, for their square is 18. If we proceed contrary to the rule by finding a surd equal to two of the surds, as $\sqrt{72}$, which is the sum of $\sqrt{32}$ and $\sqrt{8}$, and extracting the root of $10 + \sqrt{72} + \sqrt{24}$ we shall have for the two roots $\sqrt{6}$ and $\sqrt{4}$, but their square is not equal to the quantity whose root was required. The foregoing rules are illustrated by four more examples, which conclude this chapter.

CHAP. V *.

" To find the value of an unknown number, such that when it is multiplied by " a known number, and the product increased by a known number, and the sum " divided by a known number, nothing remains. Call the number by which the " unknown number is multiplied the dividend, the number which is added the " augment, and that by which the sum is divided the divisor. Find a number " which will divide these three numbers without a remainder. Perform the divi- " sion, and write the three quotients, giving each the same designation as the " number from which it was derived. Divide the dividend by the divisor, and " the divisor by the remainder of the dividend, and the remainder of the divi- " dend by the remainder of the divisor, and so on till one remains. Then let the " division be discontinued. Arrange all the quotients in a line, write the augment " below the line, and a cipher below the augment. Multiply the number above " the cipher; that is to say, the augment, by the number immediately above it, " and to the product add the cipher. Multiply the number thus found by the " number next above in the line, and to the product add the number above the " cipher, and so on till all the numbers in the line are exhausted. If of the two " numbers last found, the lower is applied according to the question, the number " above will be the quotient.

" To find the least values. Divide the value of y by a and call the remainder y. " Divide the value of x by b and call the remainder x. Multiply a by the value " of x and to the product add c. Divide the sum by b and the quotient will be " y without any remainder. And if to the first remainder we add a again and " again, and to the second remainder b as many times, we shall have new values " of x and y.

" This rule is applicable only when the number of quotients is even; when it is

---

* The rules given in this chapter are in effect the same as those which have been given by the modern European Algebraists for the solution of indeterminate problems of the first degree. Compare them with the process by continued fractions.

" odd proceed as follows. Having performed the operations directed above, sub-
" tract the value of *y* from *a* and that of *x* from *b*. If a number cannot be found
" to divide *a*, *b*, and *c*, without a remainder, but a number can be found to divide
" *a* and *c* without a remainder, (supposing the reduction of these two instead of
" that of the three which was directed by the foregoing rule) *x* will be brought
" out right and *y* wrong. To find *y* right, multiply its value now found by the
" divisor of *a* and *c*, and the product will be the true value of *y*. If *c* and *b* only
" can be reduced by a common divisor, the value of *x* must be multiplied by the
' common divisor, and the quotient will be the true value of *x*. When *c* is — sub-
" tract the value of *x* from *b*, and that of *y* from *a*."

" If the subtraction is possible let it be done, and the question is solved; if it
" is impossible suppose the excess of the subtrahend above the minuend to be
" negative. Multiply the minuend by a number, so that the product may be
" greater than the negative quantity. From this product subtract the negative
" quantity, and the remainder will be the number required.

" When *a* is — the same rule is to be observed; that is, subtract the values of *x*
" and *y* from *b* and *a*. If *c* is + and greater than *b* reject *b*, and its multiples from
" *c* till a number less than *b* remains. Note the number of times that *b* is rejected
" from *c*; if there will be no remainder after rejection it is unnecessary to reject.
" Go on with the operation, add the number of rejections to the value of *y* and
" the sum will be its true value. The value of *x* will remain as before. If *c* is —
" subtract the number of rejections from the value of *y*. If *a* and *c* are greater
" than *b* reject *b* (or its multiples) from both; call the two remainders *a* and *c*
" and proceed; *x* will come out right and *y* wrong. If there is no augment, or
" if *c* divided by *b* leaves no remainder, *x* will be = 0, and *y* the quotient. If
" the numbers are not reduced, but the quotients are taken from original num-
" bers, *x* and *y* will always be brought out right. If the numbers are reduced,
" *x* and *y* will be brought out right only when both are reduced, and but one of
" them will be brought out right when both are not reduced."

Example. $a = 221$, $c = 65$, $b = 195$, dividing these numbers by 13 we
have, $a' = 17$, $c' = 5$, $b' = 15$. Divide 17 by 15 (as above directed) continu-
ing the division till the remainder is 1. The quotients are 1 and 7, write
these in a line with c' below them, and 0 below c', thus:
Multiply 5 by 7 the product is 35, add 0 the sum is 35. Multiply 35 by 1 the product is 35, add 5 the sum is 40. The two
last numbers then are 40 and 35. From 40 throw out 17 twice,

1	40
7	35
5	
0	

6 remains; from 35 throw out 15 twice, 5 remains; therefore $x = 5$ and $y = 6$. $\frac{221 \times 5 + 65}{195} = 6$. $17 + 6 = 23$ is a new value of y, and $15 + 5 = 20$ a corresponding value of x, $2 \times 17 + 6 = 40$ is another value of y, and $2 \times 15 + 5 = 35$ a value of x. In like manner we shall have $3 \times 17 + 6 = 57$ and $3 \times 15 + 5 = 50$ new values of y and x, and so on without end.

Another Example. $a = 100$, $b = 63$, $c = 90$; c being $+$ or $-$. Although in this case 10 is a common divisor of a and c, yet as the reduction would give a wrong value of y, write a, b and c as they are, and proceed. We find the quotients 1, 1, 1, 2, 2, 1. Arrange them in a line with c below the last, and 0 below c, in this manner:

$$
\begin{array}{c}
1 \\
1 \\
1 \\
2 \\
2 \\
1 \\
\hline
90 \\
0
\end{array}
$$

We have then

$$
\begin{array}{rcl}
1 \times 90 & + 0 & = 90 \\
2 \times 90 & + 90 & = 270 \\
2 \times 270 & + 90 & = 630 \\
1 \times 630 & + 270 & = 900 \\
1 \times 900 & + 630 & = 1530 \\
1 \times 1530 & + 900 & = 2430.
\end{array}
$$

The two last numbers are 1530 and 2430, divide the former by 63 and the latter by 100; the remainders are 18 and 30, therefore $x = 18$ and $y = 30$, $\frac{100 \times 18 + 90}{63} = 30$.

By another method. Divide 100 and 90 by 10, then $a' = 13$, $b = 63$, $c' = 9$; The quotients are now found 0, 6, 3, write them in a line with c' and 0 below; we have

$$
\begin{array}{rcl}
3 \times 9 & + 0 & = 27 \\
6 \times 27 & + 9 & = 171 \\
0 \times 171 & + 27 & = 27.
\end{array}
$$

The two last numbers are 27 and 171. From 27 throw out 10 twice, 7 remains: from 171 throw out 63 twice, 45 remains. The number of quotients being odd, subtract 45 from 63, the remainder 18 is the value of x. 7 subtracted from 10 gives 3 for y, which is not the true value. To find y correct, multiply 3 by the common divisor 10, the product 30 will be the true value of y.

Another way of solving the same question is this, find a common divisor of b and c, for example, 9. Dividing b and c by 9 we have $a = 100$, $b' = 7$, $c' = 10$. Perform the division and arrange the quotients in a line with c' and 0 below, the quotients will be found 14 and 3, then

$$3 \times 10 + 0 = 30$$
$$14 \times 30 + 10 = 430.$$

From 430 throw out 100 four times, 30 remains. Here we have found a true value of y and a wrong value of x. Multiply 2 by the common divisor 9, and the product 18 is the true value of x. This question may also be solved by first taking a common divisor of a and c, and afterwards a common divisor of b and c, as follows:

Reducing a and c we have $a' = 10$, $c' = 9$, and $b = 63$. Reducing b and c we have $a = 100$, $c' = 10$, $b' = 7$. Unite the reduced numbers thus; $a' = 10$, $b' = 7$; but c having undergone two reductions *, take the difference of the numbers arising from the two operations; then $a' = 10$; $b' = 7$, $c' = 1$, divide and arrange the quotients with c' and 0, as above directed, and we shall have

$$2 \times 1 + 0 = 2$$
$$1 \times 2 + 1 = 3.$$

3 and 2 are now found for x and y, but they are both wrong, for c was reduced both with b and a. 2 must be multiplied by 9 the common divisor of b and c, and 3 must be multiplied by 10 the common divisor of a and c; the true values will be $x = 18$, $y = 30$; and new values of y and x may be had by adding a and b again and again to those already found.

---

* Let $\dfrac{ax \pm c}{b} = y$, divide a and c by p, then $\dfrac{\frac{ax}{p} \pm \frac{c}{p}}{b} = \dfrac{y}{p}$, whence $\dfrac{b\frac{y}{p} \mp \frac{c}{p}}{\frac{a}{p}} = x$, now divide b and $\dfrac{c}{p}$

by q, then $\dfrac{\frac{b}{q} \times \frac{y}{p} \mp \frac{c}{pq}}{\frac{a}{p}} = \dfrac{x}{q}$. Taking *the difference* is only true in this case, because $pq = c$, and $p - q = 1$.

What has been said is applicable only when c is $+$. When c is $-$, subtract 18, which is the value of x, from 63, the remainder is 45 ; subtract 30, which is the value of y from 100, there remains 70. We have in this case $x = 45$, $y = 70$. By adding a and b as above, new values of x and y may be found.

Another Example. Suppose $a = -60$, $b = 13$, and $c = 3 +$ or $-$. Without making any reduction, divide, and place the quotients with c and 0 as before, we have

$$1 \times 3 + 0 = 3$$
$$1 \times 3 + 3 = 6$$
$$1 \times 6 + 3 = 9$$
$$1 \times 9 + 6 = 15$$
$$4 \times 15 + 9 = 69.$$

The last numbers are 69 and 15. From 69 throw out 60, 9 remains ; from 15 throw out 13, 2 remains. The number of quotients being odd, subtract the value of x from 13, and that of y from 60, the remainders are 11 and 51. As 60 is $-$ the subtraction must be repeated, by which means we have as before $x = 2$ and $y = 9$. If c is $-$ subtract the value of x from b and that of y from a, and we shall have again $x = 11$ and $y = 51$.

Another Example. $a = 18$, $b = 11$, $c = -10$. Divide and arrange the quotients as before, we have

$$1 \times 10 + 0 = 10$$
$$1 \times 10 + 10 = 20$$
$$1 \times 20 + 10 = 30$$
$$1 \times 30 + 20 = 50.$$

From 50 reject 18, and from 30, 11 ; the remainders are 14 and 8. c being $-$ subtract 8 from 11 and 14 from 18 ; whence $x = 3$ and $y = 4$.

Another Example. $a = 5$, $b = 3$, $c = 23$. Proceeding as before, we shall have

$$1 \times 23 + 0 = 23$$
$$1 \times 23 + 23 = 46.$$

As 3 can be rejected but 7 times from 23, reject 5, 7 times from 46, the remainders are 2 and 11. If c is $-$ subtract 2 from 3 there remains 1, and 11 from 5 there remains -6. Here twice 5 must be added to -6, the sum 4 is the value of y : and that the numbers may correspond add twice 3 to 1 ; the sum 7 is the value of x, If c is greater than b, reject b from c. Throw out 3 seven

times from 23, there remains 2.   Make $c' = 2$ and place it with 0 under the line of quotients, we find $1 \times 2 + 0 = 2$
$$1 \times 2 + 2 = 4.$$

2 is the true value of x, and 4 which is found for the value of y is wrong. Add 7 the divisor of c to 4, the sum 11 is the true value of y.   If c is — subtract 2 from 3, and 4 from 5, and we shall have 1 for the value of x which is right, and 1 for the value of y which is wrong.   Subtract 7 from the value of y, the difference is — 6; add twice 5 to — 6, and we shall have 4 the true value of y.

That the numbers may correspond, twice 3 must in like manner be added to 1, and 7 will be the true value of x.

Another Example. $a = 5$, $b = 13$, $c = 0$, or $c = 65$; the quotients are 0, 2, 1, 1; place them in a line with c and 0 below, we shall have
$$1 \times 0 + 0 = 0$$
$$1 \times 0 + 0 = 0$$
$$2 \times 0 + 0 = 0$$
$$0 \times 0 + 0 = 0.$$

Add 5 to 0, which stands for the value of y, and 13 to that which stands for the value of x, we have then $y = 5$ and $x = 13$.   In the second case $a = 5$, $b = 13$, $c = 65$.   As b measures c, x will be found $= 0$ and $y = 0$.   To the value of y add 5, which is the number of times b is rejected from c, and this will give a correct value of y, for $\dfrac{5 \times 0 + 65}{13} = 5$.   Adding 13 to 0 which is the value of x, we shall have $x = 13$, and adding 5 to 5 which is the value of y, $y = 10$, for $\dfrac{5 \times 13 + 65}{13} = 10$.

Another method is to suppose $c = 1$, and proceed as above directed.   Multiply the values of x and y, which will be so found by c, rejecting a from the value of y and b from that of x, the remainders will be the numbers required.

Example. $a = 221$, $b = 195$, $c = 65$; dividing these numbers by 13, their common divisor, we have $a' = 17$, $b' = 15$, $c' = 5$.   For 5 write one, and finding the quotients as above, arrange them with 1 and 0 below, then
$$7 \times 1 + 0 = 7$$
$$1 \times 7 + 1 = 8$$

Multiply 8 and 7 by 5, the products are 40 and 35; rejecting 17 twice from 40.

and 15 twice from 35, the remainders are 6 and 5; whence $x = 5$ and $y = 6$. If c is — subtract 7 from 15 and 8 from 17, 8 and 9 remain. Multiply these numbers by 5, the products are 40 and 45; 15 and 17 being twice rejected, $x = 10$ and $y = 11$. By subtracting 6 from 17, and 5 from 15, the same numbers will be found.

" Know that the operation of the multiplicand is of use in many examples *, " as, if by the rule I shall have brought it out and any one destroys it, and some " remains; by the operation of the multiplicand, I can determine the numbers " which have been destroyed from that which remains.

" In the operation of the multiplicand of a mixed nature, the multiplicand is " of another kind, and it is called the multiplicand of addition, and that relates " to determining the value of an unknown number, which being multiplied by " a known number, and the product divided by a known number, there will re- " main a known number: and again, if the same unknown number is multiplied " by another number, and the product divided by the former divisor, the re- " mainder after division will be another number. Call the numbers by which the " unknown is multiplied the multiplicand, and that by which it is divided the " divisor, and that which is left after division the remainder. Here then are two " multiplicands, one divisor and two remainders. The method of solution is as " follows: add the two multiplicands together and call the sum the dividend. " Add the two remainders and call the sum the augment negative; leave the " divisor as it is; then proceed according to the rules which have been given: " but the values of x and y must be subtracted from b and a, x will be found " right, and y wrong."

Example. $A = 5$, $c = 7$, $a = 10$, $c = 14$, $b = 63$, $\dfrac{A.x}{b} = y + c$, and $\dfrac{a.x}{b} =$
$= z + c$, $A + a = 15$, $c + c = 21$, we have now $a' = 15$, $b' = 63$, $c' =$

---

* At this place Mr. Burrow's copy has " and besides this it is of great use in determining the signs and " minutes and seconds." And in the margin there is an example by the same commentator, apparently thus : " I give an example which comes under this rule; a star makes 37 revolutions of the heavens in 49 days and " nights; how many will it make in 17 days?" Then the writer goes on to say the answer is 12 rev., 10 s., 1°, 13′, 28″ $\frac{8}{49}$, which is got I suppose by proportion. Now he adds, " if all this were lost except $\frac{8}{49}$ it might be re- stored by the rule. He then gives the equation $\dfrac{60 \, x - 8}{49} = y$ from which x is found $= 23$ and $y = 28″$, then from $\dfrac{60 \, x' - 23}{49} = y'$ he finds x' and y', and so on till the whole is had.

—21, we find 0 and 4 $\dfrac{4 \times 7 + 0 = 28}{0 \times 28 + 7 = 7}$, $7 - 5 = 2$, $2\dot{o} - 21 = 7$, $5 - 2 = 3 = y$ wrong; $21 - 7 = 14 = x$ right; for multiplying 14 by 5 the product is 70, which being divided by 63 leaves the remainder 7; and multiplying 14 by 10, and dividing the product by 63, the remainder 14 is obtained.

CHAP. VI.

" On * the operation of multiplication of the square; and that relates to the " knowing of a square, such that when it is multiplied by a number, and to the " product a number is added, the sum will be a square.

" In this question then there are two squares, one less and the other greater, " and a multiplicand and an augment. From the multiplicand and augment " known, the two unknown squares are to be found. The method of solution " is this : Assume a number and call it the less root; take its square and mul- " tiply it by the multiplicand, and find a number which when added to it or " subtracted from it will be a square; then take its root and call it the greater " root. Write on a horizontal line these three, the less and greater roots, and the " number which was assumed as the augment. And again write such another ' line under the former so that every number may be written twice, once " above and below; then multiply crossways the two greater roots by the two " less; then take the sum of the two and call it the less root; then take the " rectangle of the two less roots and multiply it by the multiplicand, add the

---

* The rules at the beginning of this chapter for the general solution of $Ax^2 + B = y^2$ are, as they stand in the Persian, to this purport: Find $Af^2 + \beta = g^2$, where f, β, and g may be any numbers which will satisfy the equation. Make $x = fg + fg$ and $y = Aff + gg$, and $\beta' = \beta\beta$. Then $Ax'^2 + \beta' = y'^2$; and making $x'' = x'g + y'f$, and $y'' = Ax'f + y'g$, and $\beta'' = \beta'\beta$, or $x'' = x'g - y'f$ and $y'' = y'g - Ax'f$, we have $Ax''^2 + \beta'' = y''^2$. If $\beta'' \, 7 \, B$ then $\dfrac{\beta''}{p^2} = B$, and if $\beta'' \angle B$, then $\beta''p^2 = B$, but in the first case the values of x'' and y'' must be divided, and in the second case multiplied by p. In this way, by the cross multiplication of the numbers, new solutions are had for $Ax^2 + B = y^2$. When $\beta = 1$ and $\beta' = B$ the rule is the same as Fermat's proposition, which first was applied in this manner by Euler for finding new values of x and y in the equation $Ax^2 + b = y^2$. (See the investigation of this method in his algebra.) If $Ax^2 + B = y^2$, then $x = \dfrac{2r}{r^2 - A}$, r being any number: this expression is true only when $B = 1$. In that case $A \left(\dfrac{2r}{r^2 - A} \right)^2 + 1 = \left(\dfrac{r^2 + A}{r^2 - A} \right)^2$ which is the same as Lord Brouncker's solution of Fermat's problem.

" product to the rectangle of the two greater roots, the result will be the greater
" root, and the rectangle of the two augments will be the augment.

" And to find another square in the same condition write on a horizontal line
" the less root and the greater root, and the augment, which have been found,
" below the less and greater roots, and the agument which were assumed. Per-
" form the same operations as before, and what was required will be obtained.

" And another method in the operation is, after multiplying crossways to take
" the difference of the two greater roots, it will be the less root. And having
" multiplied the rectangle of the two less roots by the multiplicand, note the pro-
" duct; then take the rectangle of the two greater roots, and the difference of
" these two will be the greater root.

" And know that this augment, that is, the augment of the operation, if it is
" the same as the original augment, is what was required. Otherwise, if it is
" greater, divide it by the square of an assumed number, that the original aug-
" ment may be obtained. If it is less multiply it by the square of an assumed
" number, that the original augment may be obtained. And that they may corres-
" pond in the first case divide the greater and less roots, by that assumed number
" and in the second case multiply them by the same number.

" And a third method is this: Assume a number and divide its double by the
" difference of the multiplicand and its square, the less root will be obtained.
" And if we multiply the square of it by the multiplicand, and add the augment
" to the result, the root of the sum will be the greater root.

Example*. " What square is that which being multiplied by 8, and the pro-
" duct increased by 1, will be a square. Here then are two squares, one less and
" one greater, and 8 is the multiplicand and 1 the augment. Suppose 1 the less
" root, its square which is 1 we multiply by 8 ; it is 8. We find 1 which added
" to 8 will be a square, that is 9. Let its root which is 3 be the greater
" root. Write these three, that is to say, the less and greater roots, and the

---

* To find x and y so that $8x^2 + 1 = y^2$. Suppose $f = 1$, and $8f^2 + \beta = \square$. Let $\beta = 1$, then $8f^2 + 1 = 9 = 3^2$;
$3 \times 1 + 3 \times 1 = 6 = x$. $1 \times 1 \times 8 + 3 \times 3 = 17 = y$, $1 \times 1 = 1$ the augment; 1 being the original augment there is no
occasion to carry the operation farther. $8 \times 36 + 1 = 289 = 17^2$. For new values, $3 \times 6 + 1 \times 17 = 35 = x$,
$1 \times 6 \times 8 + 3 \times 17 = 99 = y$. $1 \times 1 = 1$ the augment. $8 \times 35^2 + 1 = 9801 = 99^2$. In like manner more values
may be found.

" augment on a horizontal line; and write these numbers below in the same
" manner, thus :

Less	Greater	Augment
1	3	1
1	3	1

" Multiply the two greater roots crossways by the two less, it is the same as it
" was before; add the two, it is 6; and this is the less root. Take the rectangle of
" the two less roots, it is 1. Multiply it by 8, it is the same 8; add it to the rect-
" angle of the two greater roots, that is 9; it is 17, and this is the greater root.
" Take the rectangle of the two augments; it is 1. As it is according to the ori-
" ginal there is no occasion to work for the original augment. The square re-
" quired is 36, which multiplied by 8 is 288; adding 1 it becomes 289, and this
" is a square whose root is 17. Again, to find a number under the same con-
" ditions. Below the less and greater roots and first augment, write the less and
" greater roots and augment which have been obtained by the operation, thus :

Less	Greater	Augment
1	3	1
6	17	1

" Multiply crossways the two greater and the two less roots, that is 3 by 6,
and 17 by 1, it is thus :

Less	Greater	Augment
1	18	1
6	17	1

" Add the two greater roots; it is 35, and this is the less root. Take the rect-
" angle of the two less roots; it is 6. Multiply it by 8, the multiplicand; add the
" product which is 48 to the rectangle of the two greater roots 3 and 17, which
" is 51, it is 99, and this is the greater root. Take the rectangle of the two
" augments; it is the original augment; for when the square of 35, which is 1225,
" is multiplied by 8, it will be 9800 ; adding 1 it will be a square, viz. 9801, the
" root of which is 99. In like manner if the two roots and the augment are
" written below the two roots and the other augment; like 6 and 17 and 1, and
" the operation is performed we shall find what we require, and another number
" will be obtained.

Another Example. " What square is that which being multiplied by 11, and
" the product increased by 1, will be a square *? Suppose 1 the less root, and
" multiply its square, which is 1 by 11; it is 11. Find a number which being
" subtracted from it, the remainder will be a square: Let the number be 2; this
" then is the negative augment, and 3 which is the root of 9 is the greater root.
" Write it thus:

Less	Greater	Augment
1	3	2
1	3	2

---

* $11x^2 + 1 = y^2$. Suppose $f = 1$ and $11f^2 - \beta = \square$. Let $\beta = 2$. $11 \times 1 - 2 = 9 = 3^2$. $3 \times 1 + 3 \times 1 = 6 = x$.
$1 \times 1 \times 11 + 3 \times 3 = 20 = y$. $-2 \times -2 = +4$. $\frac{4}{2^2} = 1$ the original augment.

Therefore $\frac{20}{2} = 10 = y$. $\frac{6}{2} = 3 = x$. For $11 \times 9 + 1 = 100 = 10^2$.

Another way. Suppose $f = 1$ and $11 f^2 + \beta = \square$. Let $\beta = 5$. $11 \times 1 + 5 = 16 = 4^2$. $4 \times 1 + 4 \times 1 = 8 = x$.
$1 \times 1 \times 11 + 4 \times 4 = 27 = y$. $5 \times 5 = 25$ the augment. $\frac{8}{5} = x$, $\frac{27}{5} = y$. $11 \times \left(\frac{8}{5}\right)^2 + 1 = \frac{729}{25} = \left(\frac{27}{5}\right)^2$.
For new values, $10 \times \frac{8}{5} + 3 \times \frac{27}{5} = \frac{161}{5} = x$. $3 \times \frac{8}{5} \times 11 + 10 \times \frac{27}{5} = \frac{534}{5} = y$. $1 \times 1 = 1$, the augment.
$11 \times \left(\frac{161}{5}\right)^2 + 1 = \frac{285156}{25} = \left(\frac{534}{5}\right)^2$. By the second method, $\frac{81}{5} - \frac{80}{5} = \frac{1}{5} = x$. $\frac{270}{5} - 11 \times \frac{24}{5} = \frac{6}{5} = y$.
1 the augment. $11 \times \left(\frac{1}{5}\right)^2 + 1 = \frac{36}{25} = \left(\frac{6}{5}\right)^2$. In the same way other values may be found.

" Multiply crossways, and add the two greater, it is 6; and this is the less
" root. Take the rectangle of the two less roots; it is 1.   Multiply by 11; it is 11.
" Add it to the rectangle of the two greater which is 9; it is 20, and this is the
" greater root.   Take the rectangle of the augments, it is 4 affirmative.   Now
" we have found a number such that when we divide this number by the square
" of that, the quotient will be the original augment.   We have found 2 and per-
" formed the operation ; 1 is obtained.   And we divide the greater root which
" is 20 by 2, 10 is the greater root.   And we divide the less root, 3 is the less
" root.   For if the square of 3 which is 9 is multiplied by 11, it will be 99, and
" when we add 1 it will be 100, and this is the square of 10 which was the
" greater root.

" Another method is, suppose 1 the less root, and multiply its square by 11,
" it is 11. We find 5 which being added to it will be a square, that is 16; its root
" which is 4 is the greater root, thus :

Less	Greater	Augment
1	4	5
1	4	5

" After multiplying crossways, add the two rectangles; it is 8, and this is the
" less root, and the rectangle of the two less ; which is 1 we multiply by 11, it is
" 11 ; add it to the rectangle of the two greater which is 16; it is 27, and this is
" the greater root.   And from the rectangle of the augments, 25 augment is
" obtained.   We have found an assumed number 5, such that when the aug-
" ment is divided by its square the quotient will be 1.   And for correspondence
" we divide 8 by 5 ; 8 fifths is the less root.   And we divide 27 by 5 ; 27-fifths
" is obtained for the greater root.   For multiplying the square of 8-fifths, that is
" 64 twenty fifth parts by 11, it is 704 twenty fifth parts ; add 1 integer that is
" 25.   It is 729 of the abovementioned denomination.   And to find other
" numbers under the same conditions, write the two roots and the other augment
" below these two roots and augment, and that is on the supposition of 3 and 10
" and 1, which were obtained before, thus :

Less	Greater	Augment
$\dfrac{8}{5}$	$\dfrac{27}{5}$	1 1
3	10	1

" After multiplying crossways, add the two rectangles, it is 161-fifths, and
" this is the less root. Multiply the less rectangle, which is 24-fifths, by 11, it is
" 264 ; add it to the rectangle of the two greater, which is 270-fifths, it is 534-
" fifths, and this is the greater root.   Take the rectangle of the augments, it is 1.
" The operation is finished, for multiplying the square of 161-fifths, which is
" 25921 twenty-fifth parts, by 11, it is 285131 twenty-fifth parts ; add 25 that
" is 1 integer, it is 285156, and this is the square of 534-fifths.

" And by the second method.   After multiplying crossways it is 81-fifths and
" 80-fifths ; the difference is 1-fifth, and this 1-fifth is the less root. Multiply the
" rectangle of the two less, which is 24-fifths, by 11, it is 264-fifths ; and the rect-
" angle of the two greater is 270 fifths.   Take the difference ; it is 6-fifths ; and
" this is the greater root ; and 1 integer is the augment.   The square of 1-fifth,
" which is 1 twenty-fifth part, multiplied by 11, is 11 twenty-fifth parts ; add 25,
" it is 36 twenty-fifth parts, the root of which is 6-fifths ; and in like manner
" any number which is wanted may be obtained.

Example.   " Let the first question be solved by the third method * : Suppose
" 3 the less root, and take the difference of its square and the multiplicand which
" is 8 ; it is 1   Divide twice 3 by 1 ; it is 6 ; and this is the less root.   For mul-
" tiplying the square of this, which is 36, by 8, it is 288 ; add 1, it is 289, and
" this is a square, the root of which is 17, and this is the greater root.

" Another method is what is called the operation of circulation †.   To bring

---

* To solve the first question by the third method.   Suppose $r = 3$, $\dfrac{2 \times 3}{9 - 8} = 6 = x$. $8 \times 36 + 1 = 289 = 17^2$. $17 = y$.

† نكوير to make to go round, from دور to go round.   This rule is, supposing $Ax^2 + B = y^2$. A and B
being given to find x and y by the operation of circulation. Find f, β and g so that $Af^2 + \beta = g^2$. Suppose

F

" out that which is required by the rule of the multiplicand. It is thus : After
" supposing the less and greater roots and the augment, suppose the less root the
" dividend, and the augment the divisor, and the greater root the augment.
" Then by the rule of the multiplicand which is passed, bring out the multipli-
" cand and the quotient. If that number by which the questioner multiplied the
" square can be subtracted from the square of this multiplicand, let it be done;
" otherwise subtract the square of this multiplicand from that number of the mul-
" tiplicand. If a small number remains, well; if not increase the multipli-
" cand thus: add the divisor again and again to the multiplicand as before
" explained, till it is so that you can subtract the number of the multipli-
" cand from the square of it, or the square of it from the number of the mul-
" tiplicand. Whatever remains we divide by the augment of the operation of
" multiplication of the square, and take the quotient which will be the augment
" of the operation of multiplication of the square. If then we shall have sub-
" tracted the multiplicand from the square, let the quotient remain as it is : and
" if we shall have subtracted the square from the multiplicand it will be contrary,
" that is, if negative it will become affirmative, and if affirmative negative ; and
" that quotient which was obtained by adding the dividend to the quotient, as
" many times as the divisor was added to the multiplicand, will be the less root ;

$\dfrac{fx + g}{\beta} = $ y and from the known numbers f, g, β, find x and y by the rules which have been given. If x^2
be $>$ A take $x^2 - $ A, or if not take A $- x^2$. If a small number remains it is well, otherwise take multiples
of β, and add them to the found value of x for a new value, till we have $(m\beta + x)^2 - $ A, or A $- (m\beta + x)^2$;
divide this by β, and if the square has been subtracted change the sign of the quotient. If instead of x the
value $m\beta + x$ has been used a corresponding value of y, $mf + $ y must be taken: by substituting these values as
follows : y, or $mf + $ y $= x'$, and $\dfrac{x^2 \backsim A}{\beta}$ or $\dfrac{(m\beta + x)^2 \backsim A}{\beta} = \beta'$, we have the solution of this equation
$Ax'^2 + \beta' = y'^2$. If β' is neither $=$ B nor to $B p^2$ nor to $\dfrac{B}{p^2}$ proceed as before. Let $Af'^2 + \beta' = g'^2$ be a solution
of $Ax'^2 + \beta' = y'^2$ where f', β' and g' are known. Suppose $\dfrac{f'x' + g'}{\beta'} = $ y'; proceed as before, and solutions
will be had for $Ax''^2 + \beta'' = y''^2$, and in like manner for $A\overset{n}{x^2} + \overset{n}{\beta} = \overset{n}{y^2}$ till $\overset{n}{\beta}$ is found $=$ B or Bp^2 or $\dfrac{B}{p^2}$.
The truth of this is plain, for as $x' = \dfrac{fx + g}{\beta}$ and $\beta' = \dfrac{x^2 - A}{\beta}$, we have $Ax'^2 + \beta' = A\left(\dfrac{fx + g}{\beta}\right)^2 + \dfrac{x^2 - A}{\beta}$ which
is $= \dfrac{(Af^2 + \beta)\, x^2 + 2Afgx + A\,(g^2 - \beta)}{\beta^2}$; but $Af^2 + \beta = g^2$, and $g^2 - \beta = Af^2$, and therefore $Ax' + \beta' =$
$\dfrac{g^2 x^2 + 2\,Afgx + A^2 f^2}{\beta^2}$ which is $= \left(\dfrac{gx + Af}{\beta}\right)^2 = y'^2$. This rule, though in some respects imperfect, is in
principle the same as that for solving the problem in integers by the application of continued fractions, which
was first given in Europe by De La Grange.

" and from the less root and the augment bring out the greater root. If then this
" augment shall have been found a square*, the operation is finished; for find
" a number by the square of which, if we divide this augment, the result will be
" the original augment, when this augment is greater than the original augment;
" or otherwise, if we multiply by it, the result will be the original augment, in
" the same manner as before. And if it is not a square perform the operation
" again in the same manner; that is, supposing the less root the dividend, and
" the augment the divisor and the greater root the augment; and work as be-
" fore till the original augment or the augment of the square is found.

Example. " What square is that which being multiplied by 67, and the pro-
" duct increased by 1, will be a square †. Let us suppose 1 the less root, mul-

---

* I suppose it should be $\text{B}p^2$ or $\frac{\text{B}}{p^2}$. I think it likely that this does not form a part of the original rule which seems to relate to integer values only.

† $67x^2 + 1 = y^2$. Suppose $f = 1$ and $\beta = -3$, then $67 \times 1^2 - 3 = \square = 64 = 8^2$, we have now $\text{A}f^2 + \beta = g^2$, where $f = 1$ and $\beta = -3$, and $g = 8$. Suppose $\frac{fx + g}{3} = \text{Y}$; that is to say, $\frac{1x + 8}{3} = \text{Y}$, reject twice β from g we have $8 - 2 \times 3 = 2$ and work for x and Y in $\frac{1x + 2}{3} = \text{Y}$. Divide 1 by 3, as directed in the last chapter. The quotient is 0, write under it 2 and 0, we find $x = 2$ and $\text{Y} = 0$. The number of quotients being odd, sub-tract the value of x from β and that of y from f. $3 - 2 = 1 = x$, $1 - 0 = 1 = \text{Y}$. As β was rejected twice from g, add 2 to the value of Y. $1 + 2 = 3 = \text{Y}$ we have now $x = 1$ and $\text{Y} = 3$. As we cannot subtract 67 from 1^2, and as a greater number will remain if we subtract 1^2 from 67, add twice β to x for a new value of x, $2 \times 3 + 1 = 7 = x$, and for a corresponding value of Y add twice f to Y. $2 \times 1 + 3 = 5 = \text{Y}$. $\text{A} - x^2$ or $67 - 7^2 = 18$. $\frac{18}{-3} = -6$. As we have taken $\text{A} - x^2$ we must change the sign of -6, it becomes $+6 = \beta'$, and $\text{Y} = x'$, we have now $\text{A}x'^2 + \beta' = y'^2$, where $\text{A} = 67$, $\beta' = 6$, and $x' = 5$, whence $y' = 41$. Since $\text{B} = 1$ and $\beta' = 6$ we pro-ceed to find $\frac{n}{\beta} = \text{B}$. Let $\text{A}f'^2 + \beta' = g'^2$ where $f' = 5$, $\beta' = 6$, and $g' = 41$. Make $\frac{5x' + 41}{6} = \text{Y}'$, we shall find $x' = 41$ and $\text{Y}' = 41$. Subtract f' 6 times from the value of Y', $41 - 6 \times 5 = 11 = \text{Y}'$, and subtract β' the same number of times from the value of x'. $41 - 6 \times 6 = 5 = x'$. $\text{A} - x'^2$ or $67 - 5^2 = 42$; $\frac{42}{6} = 7 = \beta''$. As we have taken $\text{A} - x'^2$ the sign of 7 must be changed. and $-7 = \beta''$, and $11 = \text{Y}'' = x'$; therefore $\text{A}x''^2 + \beta'' = y''^2$, and $y'' = 90$, β'' not being $= \text{B}$ we must proceed as above. Let $\text{A}f''^2 + \beta'' = g''^2$, where $f'' = 1$, $\beta'' = -7$, and $g'' = 90$. Make $\frac{11x'' + 90}{7} = \text{Y}''$, reject 7 twelve times from 90, $90 - 84 = 6$, we shall find $x'' = 12$, and $\text{Y}'' = 18$. Subtract f'' from the value of Y'' and β'' from that of x'', $18 - 11 = 7 = \text{Y}''$, $12 - 7 = 5 = x''$. The number of quotients in the division of 11 by 7 being odd, subtract the value of Y'' from f'' and that of x'' from β'', $11 - 7 = 4 = \text{Y}''$, $7 - 5 = 2 = x''$. As we cannot take 67 from 2^2, and as a greater number remains, if we subtract 2^2 from 67, add β'' once to x'' for a new value of x'', $7 + 2 = 9 = x''$; $x''^2 - \text{A}$, or $9^2 - 67 = 14$. $\frac{14}{-7} = -2 = \beta''$. As β'' was rejected 12 times from g'', 12 must be added to the value of Y, $4 + 12 = 16 = \text{Y}''$.

" tiply it by 67, it is 67. Find 3 the number of the augment, which sub-
" tracted from 67 will leave a square, that is 64, the root of which is 8, and this
' is the greater root. 1 then is the less root, and 8 the greater root, and 3 the
" augment negative. If we wish to bring it out by the operation of circulation, let
" us suppose 1 the dividend, and 8 the augment, and 3 the divisor As rejection
" of the divisor from the augment is possible, reject it twice, 2 remains. Sup-
" pose this the augment, take the numbers of the line, cipher is obtained. Write
" under it 2 the augment and cipher. Perform the operation, the multiplicand is
" found 2 and the quotient cipher. The number of the line being odd, subtract
" the multiplicand and the quotient from the divisor and the dividend, 1 and 1
" are obtained. As we rejected the divisor which is 3 from the augment which is
" 8, add 2 to the quotient, the quotient is 3 · and the multiplicand 1. As we
" cannot subtract 67 which is the multiplicand of the operation of multiplication
" of the square, from the square of this multiplicand, and if we subtract the
" square of this from 67 a greater number remains; from necessity we add the
" divisor, which is 3, twice to the multiplicand 1, it is 7; add the dividend to
" the quotient it is 5. Subtract the square of 7, which is 49, from 67, 18 re-
" mains. Divide by the augment of the operation of multiplication of the
" square, which is 3 negative, 6 negative is the quotient. As the square has
" been subtracted from the multiplicand the negative becomes contrary; it is
" 6 affirmative, and this is the augment; and 5, which was the number of
" the quotient, is the less root. Then bring out the greater root, from the less
" root and the augment, and the multiplicand 67, it is 41. Write them in order.
" As 6 is the augment of the operation and 1 is the original augment, perform the
" operation again to find the original augment: that is to say, suppose 5 the
" dividend, and 6 the divisor, and 41 the augment, and perform the operation of
" the multiplicand, the multiplicand is found 41 and the quotient also 41. Sub-
" tract 5, the dividend, 6 times from 41, the quotient, 11 remains; and subtract
" 6 the same number of times from 41, the multiplicand, 5 remains; take its

---

And as β'' was added once to the value of x'' add f'' to that of y'', $11 + 16 = 27 = y'' = x''$. Now $Af''^2 + \beta'' = g''^2$, because $x'' = 27$, and $\beta'' = -2$, therefore $y''' = 221$. Let $Af'''^2 + \beta''' = g''^2$, where $f''' = 21$, $\beta''' = -2$, and $g'' = 221$. Having now found β'', which, multiplied by itself will be the augment of the square, (meaning, I suppose, $= Bp^2$) apply the first rule of this chapter. $x''' = 2f''g''' = 11934$, $y''' = g''^2 + Af''^2 = 97684$, $\beta''^2 = 4$, we find $p = 2$ such that $\frac{\beta''}{p^2} = B = 1$. Dividing $Ax'''^2 + \beta''' = y'''^2$ by p^2, we have $\left(A\frac{x'''}{p}\right)^2 + 1 = \left(\frac{y'''}{p}\right)^2$, and $67 \times 5967^2 + 1 = 48842^2$.

" square, it is 25; subtract it from 67, 42 remains. Divide it by 6, the augment,
" 7 is the quotient, and this is the augment. As we subtracted the square of the
" multiplicand from 67, it is contrary; 7 then is the augment negative; and 11
" which is the quotient, the less root; bringing out the greater root, it is 90.
" In this case too the original augment is not obtained. Again, perform the oper-
" ation of the multiplicand; 11 is the dividend, 90 the augment, and 7 the di-
" visor. The divisor can be rejected 12 times from the augment; reject it; 6
" remains. Take the line, and perform the rest of the operation; 18 is the
" quotient and 12 the multiplicand, thus:

1	18
1	12
1	6
6	

" Subtract 11, the dividend, from 18, and 7, the divisor, from 12; 7 and 5
" are obtained, the quotient and the multiplicand. As the number of the line
" was odd subtract 7 from 11 and 5 from 7; 4 is the quotient and 2 the multi-
" plicand. As we cannot subtract 67 from the square of 2; and after subtract-
" ing the square of 2 from 67 a greater number remains, add once the divisor
" which is 7 to the dividend; it is 9. Subtract 67 from its square which is 81;
" 14 remains. Divide by 7 the augment negative, 2 negative is the quotient,
" and this is the augment. Again, as we rejected 7 twelve times from the
" augment, add 12 to the quotient which is 4; it is 16. And as we added 7 to
" 2 the multiplicand, add 11 to 16 the quotient; it is 27, and this is the less root.
" Find the greater root; it is 221. As an augment is obtained which, after being
" multiplied into itself, will be the augment of the square, we write this line
" below that, and multiply crossways in both places. From one cross multiplica-
" tion it is 5967, add these two; 11934 is obtained the less root. And the
" greater root is 97684, and the augment is 4 affirmative. We have found 2 an
" assumed number, by the square of which, if we divide this augment of the
" operation, the quotient will be 1, which is the original augment. In like

" manner we divide ¹ 1934 by 2, 5967 is the less root, and 48842 is the greater " root.

Another Example. " What square is that which being multiplied by 61, and " the product increased by 1, will be a square*. Let 1 be the less root, 8 is the " greater; and 3 the augment affirmative. Applying the operation of the multi- " plicand, it is thus :

Dividend.	Divisor.	Augment.
1	3	8

" Reject the divisor twice from the augment, 2 remains; and after the operation " 2 the multiplicand, and cipher the quotient are obtained. As the line is odd " we subtract cipher from the dividend and 2 from the divisor. It is 1 and 1. " As we rejected the divisor twice from the augment, we add 2 to the quotient. " The quotient is 3 and the multiplicand 1. If we subtract the square of the " multiplicand which is 1 from 61, a greater number remains. We therefore add " twice the dividend and the divisor to the quotient and the multiplicand. The

---

* $61 x^2 + 1 = y^2$. Let $Af^2 + \beta = g^2$, where $f = 1$, $\beta = 3$, $g = 8$. Make $\dfrac{fx + g}{\beta} = y$ that is $\dfrac{1x + 8}{3} = y$, reject β twice from g, $8 - 2 \times 3 = 2$, we shall find $x = 2$ and $y = 0$. The number of quotients in the division of 1 by 3 being odd, subtract the value of y from f, and that of x from β, $1 - 0 = 1 = y$, $2 - 1 = 1 = x$. As β was rejected twice from g add 2 to the value of y, $1 + 2 = y$. If we take $A - x^2$ a greater number remains; add twice f to the value of y, and twice β to that of x. $3 + 2 \times 1 = 5 = y$. $1 + 2 \times 3 = 7 = x$. Take $A - x^2$, $61 - 7^2 = 12$. Divide by β, $\dfrac{12}{3} = 4$, which becomes $-4 = \beta'$, and $5 = y = x'$. Now $Ax'^2 + \beta' = y'^2$, whence $y' = 39$. Let $Af'^2 + \beta' = g'^2$ where $f' = 5$, $\beta' = -4$, $g' = 39$. As β' is not $= B$, we find a number $p = 2$, such that $\dfrac{\beta'}{p^2} = -1$. Divide x' and y' by p, and we have $\dfrac{x'}{p} = \dfrac{5}{2} = x'' = f''$, and $\dfrac{y'}{p} = \dfrac{39}{2} = y'' = g''$, and $\dfrac{\beta'}{p} = -1 = \beta''$. As $B = +1$, apply the first rule of this chapter, $\beta'' \times \beta'' = -1 \times -1 = +1 = B$. $2f''g'' = 2 \times \dfrac{5}{2} \times \dfrac{39}{2} = \dfrac{390}{4} = x$, and $g''^2 + Af''^2 = \left(\dfrac{39}{2}\right)^2 + 61 \times \left(\dfrac{5}{2}\right)^2 = \dfrac{3046}{4} = y$, $\dfrac{390}{4} = \dfrac{195}{2}$, $\dfrac{3046}{4} = \dfrac{1523}{2}$ and $61 \times \left(\dfrac{195}{2}\right)^2 + 1 = \left(\dfrac{1523}{2}\right)^2$. If $Ax^2 + B = y^2$, where $B = -1$, or $61 x^2 - 1 = y^2$, then $\dfrac{5}{2} = f$, $\dfrac{39}{2} = g$, and $-1 = \beta$. Multiply $Af^2 + \beta = g^2$ crossways with $Af'^2 + \beta' = g'^2$, where $f' = \dfrac{195}{2}$ and $\beta' = +1$; in order that we may have $\beta\beta' = -1$. Then $x = 3805$, and $y = 297 8$, and $B = -1$; $61 \times (3805)^2 - 1 = (29718)^2$. For new values, where $B = +1$, multiply $Af^2 + \beta = g^2$ crossways with the values of x and y, which we have just found, $61 \times 226153980^2 + 1 = 1766319049^2$. If B is — take the product of two augments which have unlike signs, and if B is + that of two which have like signs.

" quotient is 5 and the multiplicand 7. Subtract the square of 7 from 61 ; 12
" remains. Divide by the augment of the operation of multiplication of the
" square which is 3 affirmative ; 4 affirmative is the quotient; and after reversion
" it is 4 negative ; and this is the augment; and the quotient which was 5 is the
" less root; 39 then will be the greater root. As 4 is not the original augment,
" we have found 2 an assumed number ; and by its square we divide this augment.
" 1 the augment negative is the quotient. We also divide 5 and 39 by 2. These
" same two numbers, with the denominator 2, are the quotients. As our
" question is of the augment affirmative perform the operation of cross multipli-
" cation. When we multiply the augment negative by itself it will be affirma-
" tive. The less root will be 390 fourth parts ; the greater root 3046 fourth
" parts ; and the augment 1 affirmative. Reduce the less and greater roots to
" the denominator 2. The less root is 195 second parts ; the greater root 1523
" second parts ; and the augment 1 affirmative. And if, for example, the question
" was of the subtraction of the augment, the answer would be as above ; 5
" second parts being the less root, and 39 second parts the greater root, and 1 the
" augment negative. And besides this, if we would obtain another case, let this
" be multiplied crossways with that in which 195 second parts is the less root ;
" for multiplying affirmative by negative, negative is obtained. The less root
" then is 3805, and the greater 29718, and the augment 1 negative ; and this is
" the answer to the question.

" To find another case with the augment affirmative write this below it and
" multiply crossways, 226153980 is the less root, and 1766319049 the greater
" root, and 1 the augment affirmative. And in like manner wherever the aug-
" ment is required negative, we must multiply crossways two augments of dif-
" ferent sorts ; and if affirmative two of the same sort.

Rule. " If the multiplicand of the question is the sum of two squares, and
" the augment 1 negative; it may be solved by the foregoing rules *, and if
" wished for, it may be done in another way, viz. Take the root of those two

---

* In $Ax^2 + B = y^2$, if $A = p^2 + q^2$, and $B = -1$, $x = \frac{1}{p}$ and $x = \frac{1}{q}$; for $(p^2 + q^2) \times \left(\frac{1}{p}\right)^2 - 1 = \left(\frac{q}{p}\right)^2$,
and $(p^2 + q^2) \times \left(\frac{1}{q}\right)^2 = \left(\frac{p}{q}\right)^2$

" squares, and divide the augment by each, the two numbers which are found
" will both be the less root; what was required may be obtained from each

Example. " What square is that which being multiplied by 13, when 1 is sub-
" tracted from the product, a square will remain *. 13 then is the sum of 4 and
" 9, and 1 the augment negative. Take the roots of 4 and 9, they are 2 and 3.
" Divide the augment by these two, the quotients are $\frac{1}{2}$ and $\frac{1}{3}$, both these are
" the less roots. What is required may be had from either. For multiplying
" the square of $\frac{1}{2}$ which is $\frac{1}{4}$ by 13, it is 13-fourths; and subtracting from it 1,
" which is 4, 9-fourths will remain; and this is the square of $1\frac{1}{2}$. Multiplying
" the square of $\frac{1}{3}$ which is $\frac{1}{9}$ by 13, it is 13-ninths; and subtracting 1 integer
" which is 9, $\frac{4}{9}$ remains; and this is a square."

Here follow solutions of the same question, by the former methods: I omit
them because they contain nothing new, and are full of errors in the calculation.

Another Example. Where $8x^2 - 1 = y^2$ is solved by the last rule, is omitted,
because it is immaterial.

Another Example. " What square is that which being multiplied by 6, and
" 3 added to the product, will be a square. And what number is that which
" being multiplied by 6 and 12 added to the product will be a square †. The
" operation in the first case is thus. Suppose 1 the less root, and multiply by 6,
" it is 6; add 3, it is 9; and this is a square. And for the second case thus:
" Multiply 1 by 6, it is 6; and find a number which added to it will be a square;

---

* $13x^2 - 1 = y^2$, here $A = 13 = 9 + 4 = p^2 + q^2$, $p = 2$, $q = 3$; $x = \frac{1}{2}$ and $x = \frac{1}{3}$; for $13 \times \left(\frac{1}{2}\right)^2 -$
$1 = \frac{9}{4} = (1\frac{1}{2})^2$; and $13 \times \left(\frac{1}{3}\right)^2 - 1 = \frac{4}{9} = \left(\frac{2}{3}\right)^2$.

† $6x^2 + 3 = y^2$, and $6x^2 + 12 = y^2$. First suppose $x = 1$ and $B = 3$, then $6 \times 1 + 3 = 9 = 3^2$, $y = 3$.
Second, $6 \times 1 = 6$, find β such that $6 + \beta = \square$. Let $\beta = 3$, $6 + 3 = 9 = 3^2$, 3 being not $= B$, but less
than it, find p such that $\beta p^2 = B$, $p = 2$, $3 \times 2^2 = 12 = B$. Now if $\beta' = 3$, $x' = 1$, and $y' = 3$, multiplying
$Ax'^2 + \beta = y^2$ by p^2, we have $Ax'^2 p^2 + \beta' p^2 = y'^2 p^2$, and making $x = x'p$, $y = y'p$, and $B = \beta'p^2$, we have $x = 2$,
$y = 6$ and $B = 12$. $6 \times 2^2 + 12 = 6^2$.

" we find 3. As this is not the original augment, but is less, find by the rule " which was given above, a number by the square of which when we multiply " this augment the original augment will be obtained. We have found 2. Mul- " tiply 3 by its square which is 4, it is 12; and this is the original augment. " Then that they may correspond, multiply the less and greater roots together; " also by that number, which is 2. The less root is 2, and the greater 6, and " the augment 12. Multiply the square of 2 by 6, it is 24; add 12, it is 36; " and this is a square the root of which is 6."

Here follows another example where, in $\text{A}x^2 + \text{B} = y^2$, $\text{B} = 75$, and $\text{A} = 6$. The solution of this question is like that of the first part of the preceding: f (in $\text{A}f^2 + \beta = g^2$) is assumed $= 5$ and $\beta = 75$.

Another Example. * " 300 being the augment the less root is 10; its square " which is 100, we multiply by 6, it is 600. Add 300, it is 900; and this is a " square, the root of which is 30. And know that when the augment is greater " you must bring out what you require by the operation of circulation †, that " the augment may be less. And if you wish to obtain it without the operation " of circulation call to your aid acuteness and sagacity. And when you have " found one case, and the augment is 1, you may find others without end, by " cross multiplication. For however often you multiply 1 by itself, it will still " be one; and the less root and the greater will come out different.

Rule ‡. " If the multiplicand is such that you can divide it by a square with- " out a remainder, divide it; and divide the less and greater roots by the root of " that square, another number will be found. And if you multiply it by a square " and multiply the less and greater by its root, the numbers required will also be " found.

Example. " What square is that which being multiplied by 32, and 1 added to

* $6x^2 + 300 = y^2$. Let $x = 10$, then $6 \times 10^2 + 300 = 900 = 30^2$

† When β is a greater number find β', β'', &c. less by the rule of circulation. Solutions of these problems without the rule of circulation, are to be had only by trials judiciously made.

When one case of $\text{A}f^2 + 1 = g^2$ is known any number of cases may be found by cross multiplication; for $1 \times 1 = 1$, and different values of x and y will be found at every new step.

‡ I suspect that this is incorrectly translated; the example does not illustrate the rule. Perhaps it should be, if in $\text{A}x^2 + \text{B} = y^2$, $\text{A} = \text{A}'p^2$, then $\text{A}'x^2 + \dfrac{\text{B}}{p^2} = \left(\dfrac{y}{p}\right)^2$. If $\text{A} = \dfrac{\text{A}'}{p^2}$, then $\text{A}'x^2 + \text{B}p^2 = (yp)^2$.

" the product, will be a square*.   The less root then is $\frac{1}{2}$, the square of which $\frac{1}{4}$
" multiplied by 32 will be 8 : add 1, it is 9 ; and this is a square.   If we suppose
" 2 the less root and divide 32 the multiplicand by 4, 8 will be the multiplicand ;
" and dividing the less root by the root of 4 which is 2. 1 is the less root.   For
" multiplying by 8 and adding 1, it is 9, which is a square, the root of which
" is 3.

Rule.   " If the multiplicand is a square†, divide the augment by an assumed
" number, and write the quotient in two places ; and in one place add to it, and
" in the other subtract from it the assumed number, and halve them both ; the
" greater number will be the greater root.   Divide the less by the root of the
" multiplicand, the quotient will be the less root.

Example.   " What square is that which when multiplied by 9, and 52 added
" to the product, is a square‡.   What other square is that which when multiplied
" by 4, and 33 added to the product, is a square. In the first case divide 52 by 2,
" 26 is the quotient ; write it in two places and add and subtract 2, it is 28 and
" 24 : the halves are 14 and 12 ; 14 then is the greater root.   And divide the less
" number which is 12 by the root of the multiplicand which is 3, 4 is the quo-
" tient, and this is the less root : for when the square of 4 which is 16 is multiplied
" by 9, it is 144 ; add 52, it is 196, which is the square of 14 : and in the

---

* $32 \times x^2 + 1 = \square$. Let $x = \frac{1}{2}$. $32 \times \left(\frac{1}{2}\right)^2 + 1 = 9 = 3^2$. If $x = 2$. $\frac{A}{p^2} = A'$, $\frac{32}{4} = 8$, $\frac{x}{p} = 1$. $x = 1$;
for $8 \times 1 + 1 = 3^2$.

† If $Ax^2 + B = y^2$ and $A = p^2$. take n any number; and we have $\dfrac{\frac{B}{n} + n}{2} = y$; and $\dfrac{\frac{B}{n} - n}{2} = x$; for $\dfrac{\frac{B}{n} - n}{\frac{2}{p}}$

$= \frac{B - n^2}{2pn}$ and $\dfrac{\frac{B}{n} + n}{2} = \frac{B + n^2}{2n}$. But $p^2 \times \left(\frac{B - n^2}{2np}\right)^2 + B = \left(\frac{B + n^2}{2n}\right)^2$; whence the rule.

‡ $9x^2 + 52 = y^2$ and $4x^2 + 33 = y^2$. First $\frac{52}{2} = 26$, $26 + 2 = 28$, $26 - 2 = 24$. $\frac{28}{2} = 14$; $\frac{24}{2} = 12$. $y = 14$, $\frac{12}{\sqrt{9}} = 4 = x$. $9 \times 4^2 + 52 = 196 = 14^2$. Second, $\frac{33}{3} = 11$, $11 + 3 = 14$, $11 - 3 = 8$, $\frac{14}{2} = 7 = y$, $\frac{8}{2} = 4$, $\frac{4}{2} = 2 = x$, $4 \times 2^2 + 33 = 49 = 7^2$. Values of x and y might have been found by taking $n = 4$ in the first case, and $n = 1$ in the second.

" second case divide 33 by 3, 11 is the quotient: after adding and subtracting 3 it
" is 14 and 8: after halving, the greater root is 7. Divide 4 by 2, 2 is the quo-
" tient; and this is the less root: for multiplying 4 by 4 and adding 33 to the
" product, it is 49; and this is the square of 7. And if at first we divide 52 by 4,
" and 33 by 1, what is required will be obtained

" *Another Example*, when the multiplicand is equal to the augment. What
" square is that which being multiplied by 13 and 13 subtracted from the product,
" and in another case added to it, will be a square *. In the first case suppose
" 1 the less root, its square which is also 1, we multiply by 13, it is 13: subtract
" 13, their remains cipher, the root of which is cipher. And in the second case
" suppose 3 the less root; take its square, it is 9; take the difference between it
" and the augment, 4 is the augment. Divide by it the assumed root which is
" 6, it is 6-fourths, that is $1\frac{1}{2}$, and this is the less root. The square of this
" which is 9-fourths we multiply by 13; it is 117-fourths. We see that adding
" 1 integer, that is 4-fourths to this, it is 121-fourths; and this is a square, the
" root of which is 11-second parts: the less root then is 3-second parts, and the
" greater root is 11-second parts; and the augment is 1 affirmative. As the
" original augment is 13 affirmative, perform the operation of cross multi-
" plication with the former which was 13 negative, thus: First multiply 3-second
" parts by cipher, it is cipher; and 11-second parts by 1, it is the same:

---

* $13x^2 - 13 = y^2$, and $13x^2 + 13 = y^2$. Let $x = 1$, $13 \times 1 - 13 = 0$. For $13x^2 + 13 = y^2$, let $x' = 3$, $x'^2 = 9$.
(Here are two or three errors in the Persian: A case of $Ax'^2 + 1 = y'^2$ is found by the rule $\frac{2r}{A - r^2} = x$).
$\frac{6}{4} = \frac{3}{2} = x'$. $13 \times \left(\frac{3}{2}\right)^2 + 1 = \frac{121}{4} = \left(\frac{11}{2}\right)^2$. $x' = \frac{3}{2}$, $y' = \frac{11}{2}$, and $\beta' = 1$. As $B = 13$, multiply crossways with the former case where $13 \times 1 - 13 = 0$. $\frac{3}{2} \times 0 + \frac{11}{2} \times 1 = \frac{11}{2} = x''$; $3 \times 1 \times 13 + \frac{11}{2} \times 0 =$,
$\frac{39}{2} = y''$. $13 \times - 1 = - 13 = \beta''$, but $B = + 13$. Suppose then $x''' = \frac{1}{2}$ and $\beta = - 1$, $13 \times \left(\frac{1}{2}\right)^2 - 1 = \frac{9}{4}$
$= \left(1\frac{1}{2}\right)^2$. Multiply crossways, $\frac{33}{4} + \frac{39}{4} = \frac{72}{4} = 18 = x$. And $\frac{117}{4} + \frac{143}{4} = \frac{260}{4} = 65 = y$, and $13 = B$.
Or by the rule $Aff' - gg = y$, and $f'g - fg' = x$, $\frac{39}{4} - \frac{33}{4} = \frac{6}{4} = x$, $\frac{143}{4} - \frac{117}{4} = \frac{26}{4} = y$. $13 \times \left(\frac{3}{2}\right)^2 + 13$
$= \left(6\frac{1}{2}\right)^2$.

" add them together, it is 11-second parts; and this is the less root. Multiply
" 3-second parts by 1, it is the same: multiply it by 13, the multiplicand, it is
" 39-second parts: add it to the rectangle of the two greater roots which is cipher,
" it is the same; and this is the greater root; and 13 is the augment negative;
" as it is not the original augment, for 13 affirmative is required ;   again, suppose
" the less root $\frac{1}{2}$ and the augment 1 negative; and multiply $\frac{1}{4}$ which is the square,
" by 13; it is 13-fourths. Subtract 1, that is 4-fourths, the augment negative, there
" remains 9-fourths, the root of which is $1\frac{1}{2}$.   By this we multiply crossways,
" thus :

11	39	
2	2	— 13
1	3	
2	2	— 1

" the less root is 72-fourths, which is 18 integers, and the greater root is 260-
" fourths, which is 65 integers, and the augment is 13 affirmative.
"   If we would perform the operation of cross multiplication take the dif-
" ference of the two, which are 39-fourths, and 33-fourths, that is 6 fourths ;
" $1\frac{1}{2}$ is the less root; take the difference of the two less, after multiplying by
" the multiplicand, and the rectangle of the two greater, it is 26-fourths, that
" is $6\frac{1}{2}$; and this is the greater root and 13 is the augment affirmative.

Another Example.   " What square is that which being multiplied by 5 nega-
" tive, and the product increased by 21 will be a square*. Suppose 1 the less
" root, and multiply its square by 5 negative, it is 5 negative: add 21 affirma-
" tive, it is 16 ; 4 then will be the greater root. In another way. Suppose 2 the

---

* $-5 \times x^2 + 21 = y^2$. Suppose $x = 1$; $-5 \times 1 + 21 = 16$; $y = 4$. Or, suppose $x = 2$, $-5 \times 2^2 +$
$21 = 1$, $y = 1$. By multiplying crossways when B = 1, new values may be found.

" less root and multiply its square by 5 negative, it is 20 negative; add 21
" affirmative, 1 affirmative is obtained, the root of which is 1; the less root then
" is 2, and the greater 1, and the augment 21. And if in the place of the multi-
" plicand there is 5, and the augment is 1 affirmative, multiply crossways and
" numbers without end will be obtained.

" And this which has been written is the introduction to the Indian Algebra.
" Now by the help and favour of God we will begin our object."

END OF THE INTRODUCTION.

BOOK 1.*

ON THE EQUALITY OF UNKNOWN WITH NUMBER.

———◦———

"KNOW that whatever is not known in the question, and it is required to bring
" it out by a method of calculation, suppose the required number to be one
" or two unknown, and with it whatever the conditions of the question in-
" volve, and proceed by multiplication and division, and four proportionals and
" five proportionals, and the series of natural numbers, and the knowledge of the
" side from the diameter, and the diameter from the side, that is the figure of the
" bride†, and the knowledge of the perpendicular from the side of the triangle,
" and conversely, and the like, so that at last the two may be brought to equa-
" lity. If after the operation they are not equal, the question not being about
" the equality of the two sides, make them equal by rejection and perfection, and
" make them equal. And that is so, that the unknown, and the square of the
" unknown of one side is to be subtracted from the other side, if there is an un-
" known in it; if not subtract it from cipher: and subtract the numbers and
" surds of the other side from the first side, so that the unknown may remain
" on one side, and number on the other; the number then, and whatever else is
" found, is to be divided by the unknown, the quotient will be the quantity of
" the unknown.

" If the question involves more unknown quantities than one, call the first
" one unknown, the second two unknown, the third three unknown, and so on.
" And the method is this. Suppose the quantity of the lower species less than
" that of the higher, and sometimes suppose $\frac{1}{2}$, and $\frac{1}{3}$, and $\frac{1}{4}$ of the unknown

---

* There are many parts of the rules given in the rest of the Work which are unintelligible to me; they
are obscured probably by the errors of transcribers and of the Persian translator.—I translate them as exactly
as I can from the Persian.

† The Arabs call the 47th proposition of the first book of Euclid, " the figure of the bride." I do not
know why.

" and the like : and sometimes suppose the unknown to be a certain number, and
" sometimes suppose 1 unknown and the rest certain numbers." The shortest
method of solving the question is directed to be observed, and the whole attention to be given to what is required.

The first example is, " A person has 300 rupees and 6 horses; and another
" person has 10 horses and 100 rupees debt; and the property of the two is
" equal; and the price of the horses is the same; what then is the value of each?
" Or, the first person has two rupees more than the property of the first person
" in the first question, that is 3 horses and 152 rupees; and the second has the
" same as he had before, and the property of both is equal; what then is the
" price of one of the horses? Or, in the first question, the property of the first
" person is three times the value of that of the second, what then is the value of one
" horse? The operation in the first question is this: I suppose the price of a horse
" to be the unknown; 6 horses are six unknown. The first person's property then
" is 300 rupees affirmative and 6 unknown : and the property of the second is 10
" unknown and 100 rupees negative. As by the question both these sides are
" equal there is no occasion for the operation of rejection and perfection. I
" make them equal in this manner:

+ 300 Rupees	$6x$
− 100 Rupees	$10x$

" First I write them both, above and below, and I take 100 rupees negative from
" 300 rupees affirmative, it is 400 rupees affirmative; and I take 6 unknown from
" 10 unknown, there remains 4 unknown. 400 rupees is equal to 4 unknown. I
" divide the first by the second, 100 rupees is the quotient, and this is the price
of a horse." The other questions produce also simple equations, in which nothing
remarkable occurs.

The second example has three unknown quantities with only one equation; it
is solved first by assuming the unknown quantities in the proportion of 3, 2, and
1; and secondly by assuming them as 1, 5, and 4.

In the third example the Mussulman names, Zeid and Omar are introduced.
The fourth and fifth examples contain nothing worthy of notice.

The sixth is as follows: " A person lent money to another on condition that he

" should receive 5 per cent. a month.   After some months he took from him the
" principal and interest, and having subtracted the square of the interest from
" principal gave the remainder to another person, on condition that he should
" receive 10 per cent. and after the same time had passed, as in the former case,
" he took back the principal and interest, and this interest was equal to the first
" interest; what sum did he lend to each person, and what was the time for
" which the money was lent*[2]"

The first principal is supposed unknown, and the number of months during
which it was lent is supposed 5.   The question is solved by the rules of proportion and a simple equation.   Another way is given for working this question, viz.

" Divide the interest of the second by that of the first, call the quotient the
" multiplicand, and suppose a number the interest for the whole time and take its
" square, and from the multiplicand subtract 1, and divide the square by the
" remainder; the quotient will be the amount of the second sum, and the second
" sum multiplied by the multiplicand, or added to the square of the interest of
" the whole, will be equal to the first sum."

The next question is like the preceding, and is solved by means of the rule.
I pass over several other examples, which contain nothing new or remarkable.
A question in mensuration comes next.

" There is a triangle, one side of which is 13 surd, and another side 5 surd,
" and its area 5 direhs; how much is the third side? I suppose the third side
" unknown; the side 13 is the base.   It is known that when the perpendicular
" is multiplied by half the base, or the base by half the perpendicular, the pro-
" duct will be the area of the triangle.   Here the base and the area are known,
" and the perpendicular is unknown: I divide 4 which is the whole area by half
" of 13 surd; the quotient is the perpendicular.   I perform the operation thus:
" As 4 is a number I take its square 16, for the division of a number by a surd
" is impossible.   I take half of 13 surd thus: I square 2, which is the denomi-
" nator of $\frac{1}{2}$, it is 4.   I divide 13 by it.   The quotient is 13 parts of 4 parts.   I

---

* Let P, p, be the principal, I, i, the interest; R, r, the rate, and N, n, the number of the months. If
$prn = i$, PRN = I, $P = p - i^2$, N = n, and I = i; we have $(p - i^2)$. RN = $i = prn$; or $pRN - i^2RN = prn$; whence $pn \times (R - r) = i^2RN$, and $R - r = \dfrac{i^2 RN}{pn}$, but this is equal to $\dfrac{ri^2}{P}$, and $P = \dfrac{ri^2}{R - r} = \dfrac{i^2}{\dfrac{R}{r} - 1}$,

which is the first part of the rule; the rest is evident.

" divide 16 by 13 parts of 4 parts ; it is 64 parts of 13 surd; and this is the
" perpendicular. I then require the excess of the square of 5 surd above 64 parts
" of 13 surd : First I take the square of 5 surd ; it is 5 number ; take its square,
" it is 25 surd ; the root of which is 5. I then take the square of 64 parts of 13
" surd, as above. I take the excess thus : I make 5 of the same sort ; it is 65 ;
" I take the excess of 65 above 64 ; it is one part of 13 surd ; and this is from
" the place of the perpendicular to the angle formed by the side 5 and the base."

The other segment of the base is found by subtracting this from the whole, by
a rule which was given in the 4th chapter of the introduction, for finding the
difference of two surds, viz. $\sqrt{a}-\sqrt{b}=\sqrt{\left(\sqrt{\left(\frac{a}{b}-1\right)^2} \times b\right)}$. The square root
of the sum of the squares of this segment and the perpendicular gives the quantity
of the unknown side of the triangle.

In the next question, the sides of a triangle being given, its area is required.
One of the segments of the base made by a perpendicular, is supposed unknown.
From two values of the perpendicular, in terms of the hypothenuses of the two
right-angled triangles, and their bases, an equation is formed ; from which the
unknown quantity is brought out. The equation involves many surds, and they
are reduced by the rules laid down in the introduction. The perpendicular is then
found by taking the square root of the difference of the squares of a segment of
the base, and of the adjacent sides of the triangle. The operation is here con-
cluded. In a marginal note are directions to find the area, as in the foregoing
case.

The next is, " What four fractions are those whose denominators are equal, and
" whose sum is equal to the sum of their squares. Also what four fractions are
" those, the sum of whose squares is equal to the sum of their cubes." For the
first part of the question : " Suppose the first fraction one unknown, the second
" two unknown, the third three unknown, and the fourth four unknown, and
" below each write 1 for the denominator. The sum of the four is 10 unknown.
" Their squares are 1 and 4 and 9 and 16, whose sum is 30 square of unknown,
" and these two quantities are equal. Divide both by one unknown ; the quotients
" are 10 number and 30 unknown. Divide 10 by 30 unknown, the quotient is
" $\frac{1}{3}$ of unknown. The first fraction then is $\frac{1}{3}$, the second $\frac{2}{3}$, the third $\frac{3}{3}$, and
" the fourth $\frac{4}{3}$; and the squares of these fractions are $\frac{1}{9}$ and $\frac{4}{9}$, and $\frac{9}{9}$ and $\frac{16}{9}$; and

" the sum of these four is $\frac{30}{9}$, and this is equal to $\frac{10}{3}$." In the same manner the the other fractions are found to be $\frac{3}{10}$, $\frac{6}{10}$, $\frac{9}{10}$, and $\frac{12}{10}$.

The next is to find a right-angled triangle, " the area of which is equal to its " hypothenuse;" and to find a right-angled triangle, " the area of which is equal " to the rectangle of its three sides." For the first part of the problem, one side of the triangle is assumed equal to 4 unknown, and the other side equal to 3 unknown; the hypothenuse is found equal to $5x$, and the area equal to $6x^2$; the equation $5x = 6x^2$ being reduced, gives the value of x. For the second part, the sides are assumed as above, and the value of x is deduced from the equation $60x^3 = 6x^2$.

The next problem is, to find two numbers of which the sum and the difference shall be squares, and the product a cube. The numbers are supposed $5x^2$ and $4x^2$, and, the cube to which their product must be equal $1000x^3$, whence x is found.

The next is to find two numbers such that the sum of their cubes shall be a square, and the sum of their squares a cube. One number is supposed x^2, and the other $2x^2$, and the cube $125x^3$ *. In the solution of this the following passage occurs: " The cube of the square of unknown, which in Persian algebra " is termed square of cube." In the margin is this note: " Here is evidently a " mistake; for in Persian algebra the unknown (مجهول) is called thing (شیٔ), " and its square (مربع) square (مال), (literally possession;) and its cube " (مكعب) cube (كعب); and when the cube is multiplied by thing, the " product is called square of square (مال مال); and when the square of " square is multiplied by thing, the product is called square of cube (مال كعب); " and when the square of cube is multiplied by thing, the product is called cube " of cube (كعب كعب), not square of cube. For example, suppose 2 thing " 4 is its square, 8 its cube, 16 its square of square, 32 its square of cube, 64 its " cube of cube, not its square of cube, although it is the cube of the square " (مكعب مربع), or the square of the cube (مربع مكعب)."

In the next example the three sides of a triangle are given, and the perpen-

---

* Sum of the cubes $= x^6 + 8x^6 = 9x^6$ (a square); and the sum of squares $= x^4 + 4x^4 = 5x^4$, assume this $= 125x^3$, or $5x^4 = 125x^3$, whence $5x = 125$, and $x = 25$; therefore 625 and 1250 are the numbers.

dicular is required. It is found in the same way as the perpendicular was found in one of the former questions, when the sides being given the area of the triangle was required.

The three following are different cases of right-angled triangles in which the parts required are found by the principle of the square of the hypothenuse being equal to the sum of the squares of the two sides, and simple equations. In the first the base and the sum of the hypothenuse and the other side are given. In the second the base and the difference of the hyptohenuse and the other side are given; and in the third the base, part of one side, and the sum of the hypothenuse and the other part of that side, are given.

The first book ends with the following example: " Two sticks stand upright " in the ground, one is 10 direhs in height and the other 15 direhs, and the " distance between the two is 20 direhs. If two diameters are drawn between " them, what will be the distance from the place where they meet to the ground *? " Suppose the perpendicular unknown; it is known that as 15 to 20, so is the " unknown to the quantity of the distance from the side 10 to the place where the " unknown stands. We find then by 4 proportionals, 4 thirds of unknown is the said " quantity. In like manner we find the second quantity 20 parts of 10, that is " 2 unknown. Take the sum of the two, it is 10-thirds, and this is equal to 20. " Divide 20 by 10-thirds, the quotient is 6; and this is the quantity of the un- " known, that is of the perpendicular. From the place where the perpendicular " stands on the ground, to the bottom of the side 15, is 12; for it is 2 unknown. " The second quantity is 8 direhs; for it is 1 unknown and a third of unknown. " And know that whatever the distance is between the two sticks, the quantity " of the perpendicular will be the same; and so it is in every case. We can also

---

* Let $AB = 10$, $DC = 15$, $BD = 20$.

By similar triangles $BD : BP :: DC : PG$,

$BD : PD :: BA : PG$,

whence $BP : PD :: BA : DC$,

therefore BD is divided in P in the ratio of DC to BA.

By composition $BP + PD : BP :: BA + DC : BA$; but $BP + PD = BD$, therefore $BA + DC$ and BA are in the ratio of BD to BP; whence, by the first proportion, $BA + DC : BA :: DC : PG$, that is, PG is a fourth proportional to $BA + DC$, BA, and DC, whatever be the length of BD.

Lucas de Burgo has this proposition, (see his Geometry, p. 56.) where the lengths are 4, 6, and 8; or page 60, where they are 10, 15, and 6. The same is in Fyzee's Lilavati, where the rules are

$$GP = \frac{AB \times CD}{AB + CD}, \quad PD = \frac{BD \times CD}{AB + CD}, \text{ and } BP = \frac{BD \times AB}{AB + CD}.$$

" ascertain these two quantities by another method, and that is the ratio of 25,
" (that is the sum of the two sides) to 20, is like the ratio of 15 to the unknown ;
" that is the quantity towards the side 15. Multiply 15 by 20, it is 300. Divide
" 300 by 25, it is 12. The ratio of 25 to 20, is like the ratio of 10 to the unknown ;
" the result is 8, and this is the quantity towards the side 10. By another
" method, by four proportionals, we find that the ratio of 20 to 25, is like the
" ratio of 8 to the unknown ; 6 is the result. In like manner the ratio of 20 to
" 10, is like that of 12 to unknown ; again 6 is the result. Another method is,
" divide the rectangle of the two sticks by the sum of the two, the result is the
" quantity of the perpendicular, and the quantity of the ground we multiply by
" each side separately, and divide both by the sum of the sides. The two quotients
" will be the quantities from the place of the perpendicular to the bottom of the
" sticks ; accordingly divide 150, which is the rectangle of the two sticks, by
" 25, the quotient is 6. Multiply 20 dirchs, which is the quantity of the ground
" by both sticks, the products are 300 and 200 Divide both by 25, the quotients
" are 12 and 8. In this manner the figure may be found by calculation as
" correctly as if it were measured."

END OF THE FIRST BOOK.

BOOK 2.

———◦◦✕◦◦———

" On the interposition of the unknown : where the square of unknown is
" equal to number, and that is rejected with the unknown." *(Or divided by
the unknown* (ببجهول ردكتد ازا). *(I do not know what he means here,
perhaps there is some error.)*

" It is intitled, 'Interposition of Unknown,' (توسيط مجهول); because
" that which is required is brought out by means (واسط) of the unknown. It is
" called Interposition (توسيط); and Mudhum Uhrun (unknown means) in
" Hindee is to be so understood. Its method is this : The square of unknown
" being equal to number, multiply both, or divide both by an assumed number,
" and add a number to the two results, or subtract it from them that both may
" be squares. For if one side is a square the other also will be a square ; for they
" are equal, and by the equal increase or diminution of two equals, two equals
" will be obtained. Take the roots of both, and after equating divide the num-
" ber by the root of the square of unknown, that is by unknown ; the result will
" be what was required. And if there is equality in the cube of the unknown,
" or the square of the square, that is after the operation in the thing and cube,
" and square of square, if the root cannot be found, nor be brought out by rule,
" in that case it can only be obtained by perfect meditation and acuteness *.
" And, after equating, if the two sides are not squares, the method of making
" them squares is this. Assume the number 4, and multiply it by the number of
" the square of the first side, and multiply both sides by the product. And in the

---

* From this place to the end of the rule Mr. Burrow's copy is as follows : " And if in the side which has the
" unkno vn there is a number greater than the unknown, if the number is affirmative make it negative, and if
" negative, two numbers will be found in the conditions required, and the way to find the assumed number by
" which the two sides should be multiplied, and the number to be added, is extremely easy ; for multiply the
" multiplicand of the number of the square of the unknown by 4, and let the square of the numbers of the un-
" known of the side in which there is the square of the unknown, be the number added."

" place of number increase both sides by the square of the thing of the unknown,
" which is on that side; both sides will be squares. Take the roots of both and
" equate them, and the quantity of the unknown will be found.

Example. " Some bees were sitting on a tree; at once the square root of half
" their number flew away. Again, eight-ninths of the whole flew away the
" second time; two bees remained. How many were there? The method
" of bringing it out is this: From the question it appears that half the sum has
" a root; I therefore suppose 2 square of unknown, and I take 1 unknown, that
" is the root of half. And as the questioner mentions that two bees remain, 1
" unknown and $\frac{8}{9}$ of 2 square of unknown, that is $\frac{16}{9}$ of 1 square of unknown,
" and 2 units, is equal to 2 square of unknown. I perform the operation of
" equating the fractions in this manner, I multiply both sides by 9, which is the
" denominator of a ninth; 16 square of unknown and 9 unknown, and 18 units,
" is equal to 18 square of unknown. I equate them thus: I subtract 16 square of
" unknown of the first side from 18 square of unknown of the second side; it is
" 2 square of unknown affirmative; and in like manner I subtract 9 unknown of
" the first side from cipher unknown of the second side; 9 unknown negative
" remains. Then I subtract cipher the numbers of the second side from 18 units of
" the first side; it is the same. The first side then is 2 square of unknown affirma-
" tive and 9 unknown negative, and the second side is 18 units affirmative. In this
" example there is equality of square of unknown, and unknown to number; that
" is equality of square and thing to number. As the roots of these two sides can-
" not be found, suppose the number 4, and multiply it by 2, which is the number
" of the square of the unknown, it is 8. I multiply both sides by 8; the first side is
" 16 square of thing, and 72 unknown negative; and the second side is 144
" units.   I then add the square of the number of the unknown, which is 81, to the
" result of both sides; the first side is 16 square of unknown, and 72 unknown
" negative, and 81 units; and the second side is 225 units. I take the roots of
" both sides: the root of the first side is 4 unknown and 9 units negative; and
" the root of the second side is 15 units affirmative. I equate them in this
" manner: I subtract cipher unknown of the second side from 4 unknown of the
" first side; and 9 units negative of the first side from 15 units affirmative of
" the second side; the first is 4 thing, and the second side is 24 units affirma-
" tive, I divide, 6 is the result, and this is the quantity of the unknown; and

" as we supposed 2 square of unknown, we double 36 ; the whole number of bees
" then was 72."

In the next example arises the equation $\frac{x^2}{2} + 4x + 10 = x^2$; then $x^2 - 8x = 20$,

$4x^2 - 32x + 64 = 4 \times 20 + 64 = 144$, $2x - 8 = 12$; $x = 10$[*].

The next example is, " A person gave charity several days, increasing the
" gift equally every day. From the sum of the days 1 being subtracted and
" the remainder halved, the result is the number of dirhems which he gave the
" first day ; he increased always by half of that number: and the sum of the
" dirhems is equal to the product of these three ; that is to say, the number of
" days, the number of dirhems the first day, and the number of the increase,
" added to $\frac{1}{7}$ of the product." Let the number of days be $4x + 1$, $2x$ is the
" number of dirhems given the first day, and x is the number of the increase ;

$(4x + 1) \times 2x \times x = 2x^2 + 8x^3$, add $\frac{1}{7}$ of this $\frac{16x^2 + 64x^3}{7}$ = the sum of the

dirhems. Then by a rule of the Leelawuttee, $((4x + 1) - 1) x + 2x$ = the
number given the last day ; half the sum of what was given the first and
last days = what was given the middle day. Multiply this by the number of

days ; it is $8x^3 + 10x^2 + 2x$, which is $= \frac{16x^2 + 64x^3}{7}$. $8x^2 - 54x = 14$. Mul-

tiply by the assumed number 8, for in this case as the co-efficient of x is even,
assume the co-efficient itself of x^2, and add the square of half the co-efficient of

x, $64x^2 - 432x + 729 = 841$; $8x - 27 = 29$, $x = \frac{29 + 27}{8} = 7$.

The next is to find x in the equation $\left(\frac{x^2}{0} + \frac{x}{0}\right) \times 0 = 90$. The solution is,

" I suppose, what is required to be *thing*, I divide it by cipher; as the quotient
" is impossible, *thing* is obtained, whose denominator is cipher. Its square, which
" is the square of *thing*, whose denominator is cipher, I add to *thing*, which is the
" root. It is the square of *thing* and *thing*, whose denominator is cipher. I
" multiply by cipher; it is the square of *thing* and *thing*. Cipher is thrown out
" by a rule of the Leelawuttee, which says, that when the multiplicand is cipher,
" and the multiplier a number whose denominator is cipher, the product will be

---

* Part of this example, and most of the rest in this book, are wanting in Mr. Burrow's copy.

" that number, and cipher will be rejected." Whence the equation $x^2 + x = 90$ which is solved in the common way.

The next is; a value of x is required in the case $\left(\left(x + \frac{x}{2}\right) \times 0\right)^2 + 2 \left(x + \frac{v}{2}\right) \times 0 = 15$. It is brought out in a manner similar to that of the foregoing.

The next example is of a cubic equation, viz. $x^3 + 12x = 6x^2 + 35$. The terms involving the unknown quantity being brought all on the same side, 8 is added to complete the cube. " I take the cube root of the second side 3, and I " write the terms of the first side in the arithmetical manner, thus: 8 units nega- " tive, and 12 unknown affirmative, and 6 square of unknown negative, and 1 " cube of unknown. First, I take the cube root of the last term, it is 1 un- " known. I square it and multiply it by 3, and I divide the term which is last " but one by the product ; 2 units negative is the quotient. Its square, which is 4 " affirmative, I multiply by the term first found, viz. 1 *thing* ; it is 4 *thing*. I " multiply it by 3, it is 12 unknown. I subtract it from the third term which is " after the first, nothing remains. After that I subtract the term 2 negative from " the first term, nothing remains. The cube root then of the first side is found 1 " *thing* affirmative and 2 units negative." Whence $x - 2 = 3$, which is re- duced in the usual way.

In the next a biquadratic is found, $x^4 - 400x - 2x^2 = 9999$. To solve this $400x + 1$ is directed to be added to each side ; the equation is then $x^4 - 2x^2 + 1 = 10,000 + 400x$. The root of the first side is $x^2 - 1$, but the root of the second side cannot be found. Find a number which being added, the roots of both sides may be found; that is $4x^2 + 400x + 1$. This will give $x^4 + 2x^2 + 1 = 10,000 + 4x^2 + 400x$; and extracting the square root, $x^2 + 1 = 100 + 2x$, which is reduced by the rules given in this chapter. At the conclusion of the ex- ample are these words : " The solution of such questions as these depends on " correct judgment, aided by the assistance of God.

In the two next examples notice is taken of a quadratic equation having two roots. " When on one side is *thing*, and the numbers are negative, and on the " other side the numbers are less than the negative numbers on the first side, " there are two methods. The first is, to equate them without alteration. The " second is, if the numbers of the second side are affirmative, to make them nega- " tive, and if negative to make them affirmative. Equate them ; 2 numbers will " be obtained, both of which will probably answer."

The next example is, " The style of a dial 12 fingers long stands perpendi-

" cular on the ground. If from its shadow, a third of the hypothenuse of these
" two sides, viz. the style and the shadow, is subtracted, 14 fingers will remain.
" What then are the shadow and the hypothenuse?" In the right-angled

triangle $a\angle\!\!\!\!\diagdown \,^c_b$, a being $= 12$, and $b - \dfrac{c}{3} = 14$; c and b are required. The equa-

tion $(3b - 42)^2 = b^2 + 144$ arises, and is reduced to $4b - 63 = 27$. The two

values $\pm\ 27$ are taken notice of. First, $+\ 27$ gives $b = 22\dfrac{1}{2}$, which is

declared to be right. From $-\ 27$, b is found $= 9$; " but here," it is observed,
" 9 is not correct; for, after subtracting a third of the hypothenuse, 14 does
" not remain." In opposition to this, some one speaking in the first person (the
Persian translator, I suppose) says, " I think that this also is right," and goes on
" to prove that in this case the hypothenuse will be $= -\ 15$."

The next problem is to find four numbers such that if to each of them 2 be
added, the sums shall be four square numbers whose roots shall be in arithmetical
progression; and if to the product of the first and second, and to the product of
the second and third, and to the product of the third and fourth, 18 be added,
these three sums shall be square numbers; and if to the sum of the roots of all
the square numbers 11 be added, the sum shall be a square number, viz. the
square of 13.

It is here observed, by way of lemma*, that, in questions like this, the "aug-
" ment of the products" must be equal to the square of the difference of the roots,
multiplied by the " augments of the numbers;" otherwise the case will be impossible.

The following is an abstract of the solution: (Let w, x, y, z be the four numbers required, and r, s, t, v the four roots which must be in arithmetical pro-

gression). By the lemma we find the common difference $\sqrt{\dfrac{18}{2}} = 3$. The first

root being r, the second will be $= r + 3$, the third $= r + 6$, and the fourth $=$
$r + 9$.

Now $rs - 2 = \sqrt{(wx + 18)}$, and $st - 2 = \sqrt{(xy + 18)}$, and $tv - 2 = \sqrt{(yz + 18)}$.

---

* In a marginal note, which I suppose to be written by the Persian translator, the application of the Lemma
to the problem is illustrated thus: Let a, b, c be three numbers: $(a - b)^2 \times c + (a^2 - c) \times (b^2 - c) = (ab - c)^2$.
In this case we have $a - b = 3$, $c = 2$ and b and a two successive roots; and as $w = r^2 - 2$, and $x = s^2 - 2$, &c.
the reason of the rule is plain.

We have now $w+2=r^2$; $x+2=(r+3)^2$; $y+2=(r+6)^2$; $z+2=(r+9)^2$; And $wr+18=(rs-2)^2$; $xy+18=(st-2)^2$, and $yz+18=(tv-2)^2$.

Making $r+s+t+v+(rs-2)+(st-2)+(tv-2)+11=13^2$; a quadratic equation arises, which being reduced r is found $=2$, whence $w=2$, $x=23$, $y=62$, and $z=119$.

Some questions about right-angled triangles occur next; the first is, " Given " the sides of a right-angled triangle 15 and 20; required the hypothenuse. " Although by the figure of the bride the hypothenuse is the root of the sum of " the squares of the two sides, the method of solution by lgebra is this : In this " triangle suppose the hypothenuse unknown, and then divide the triangle into " two right-angled triangles, thus : Suppose the unknown hypothenuse the base " of the triangle, and from the right-angle draw a perpendicular ; then 15 is the " hypothenuse of the small triangle, and 20 that of the large one. By four pro- " portionals I find, when the least side about the right angle, whose hypothe- " nuse is 1 unknown, is 15; how much will be the least side about the right " angle whose hypothenuse is 15."   In like manner the other segment is to be brought out, whence $x=25$. " If I would find the quantity of the perpen- " dicular, and the segments of the hypothenuse at the place of the perpendicular, " it may be done in various ways ; first by four proportionals," &c. They are found on the same principle as above. " And another way which is written " in the Leelawuttee is this ; The difference of the two containing sides, that is " to say 5, I multiply by 35, which is the sum of the two sides; it is 175. " I divide by 25 that is the base; the quotient is 7. I add this to the base, it " is 32.   I halve it, 16 is obtained ; when I subtract 7 from the base, 18 " remains. I halve, 9 is the smaller segment from the place of the perpen- " dicular.

Rule.   " The square of the hypothenuse of every right-angled triangle is equal " to twice the rectangle of the two sides containing the right angle, with the " square of the difference of those sides. As the joining of the four triangles " abovementioned is in such a manner that from the hypothenuse of each, the " sides of a square will be formed, and in the middle of it there will be a square, " the quantity of whose sides is equal to the difference of the two sides about the " right-angle of the triangle ; and the area of every right-angled triangle is half " the rectangle of the sides about the right triangle. Now twice the rectangle " of the two sides containing *that* is 600, is equal to all the four triangles ; and

" when I add 25, the small square, it will be equal to the whole square of the
" hypothenuse, that is 625, which is equal to the square of thing ; and in many
" cases an effable root cannot be found, then it will be a surd ; and if we do
" not suppose thing, add twice the rectangle of one side into the other, to the
" square of the difference of the sides, and take the root of the sum, it will be
" the quantity of the hypothenuse. And from this it is known that if twice the
" rectangle of two numbers is added to the square of their difference, the result
" will be equal to the sum of the squares of those two numbers."

The next is in a right-angled triangle . Given $\sqrt{(AB - 3)} - 1 = AC -$
BC, required the sides. " First, I perform the operation of contrariety and op-
" position: let AC — BC be supposed 2. To this add 1, it is 3; take its square,
" its is 9 ; add 3, it is 12. This is the quantity of the less side ; its square which
" is 144 is $= AC^2 — BC^2$; here then the differences of the two original numbers,
" and of the two squares are both known ; and the difference of the squares of two
" numbers is equal to the rectangle of the sum of the two numbers, into their
" difference. Therefore when we divide the difference of the squares by the
" difference of the two numbers, the sum of the two numbers will be the quotient;
" and if we divide by the sum, the difference will be the quotient: because the
" square of a line has reference to a four-sided equiangular figure whose four
" sides are equal to that line ; for example, the square of 7 direhs is 49. If I
" subtract the square of 5 from it, 24 remains; and the difference of 7 and 5 is
" 2, and their sum 12, and the rectangle of these two is 24, which is the number
" remaining. Then it is known that the rectangle of the sum of the two numbers
" into their difference, that is 12 multiplied by 2, is equal to the difference of the
" squares of the two that is 24,", &c. On this principle the sum and difference
being found, the numbers themselves are had " by a rule of the Leelawuttee," viz.

$$\frac{a + b}{2} + \frac{a - d}{2} = a, \text{ and } \frac{a + b}{2} - \frac{a - b}{2} = b.$$

By supposing other numbers besides 2 for the difference, and proceeding in the
above manner, triangles without end may be found.

As objection is here made (I suppose by the Persian translator) that the
above is not algebraical. It is then stated that the translator has found out an
easy way of solving the question by Algebra. He directs that the difference
AC — BC may be assumed = 2, as before ; and making BC = x, AC will be

$= x + 2$, and AB being $= 12$, the value of x may be found from the equation $x^2 + 12^2 = (x + 2)^2$.

Rule. " The difference of the sum of the squares of two numbers and the
" square of their sum is equal to twice the rectangle of the two numbers. For
" example, the squares of 3 and 5 are 9 and 25, that is 34, and their sum
" is 8 and its square 64, and the difference of these is 30, which is equal to twice
" the rectangle of 3 and 5 that is by the 4th figure of the second book thus." In
the copy which I now have, the figures are omitted. In Mr. Burrow's copy it is

Then follows another rule: $4ab - (a + b)^2 = (a - b)^2$, which may be easily
understood by this figure. There is no figure in Mr. Burrow's copy, nor in my
present copy, but I had one in which there was a figure for the demonstration of
the 8th proposition of the second book of Euclid.

Next come two examples : The first of them is, what right-angled triangle is
that " the sum of whose 3 sides is 40, and the rectangle of the two sides about
" the right angle 120 ?

" The method of solution is this : By the first rule take twice 120, it is 240,
" and this is equal to the difference of the sum of the squares of the sides about
" the right angle, and the square of their sum that is the hypothenuse. Then
" the difference of the squares of the two numbers, one of which is the sum of the
" two sides and the other the hypothenuse, is 240 ; and the sum of both is 40.
" In the method of finding out the triangle, it was before known that the dif-
" ference of the squares of two numbers is equal to the rectangle of their sum and
" difference ; when the difference of the two squares is divided by the differ-
" ence of the two numbers, the quotient is the sum of the numbers ; and if it is
" divided by the sum, the quotient is the difference. Let then 240 be divided
" by the two numbers, which together make 40 by the question, the quotient

" is 6, and this is the difference of the hypothenuse and the sum of the two sides
" about the right angle; then add 6 to 40, and take its half, it is 23; this is the
" sum of the sides; subtract 6 from 40, and take its half, it is 17, and this is the
" hypothenuse, for the sum of the two sides is always greater than the hypothe-
" nuse by the asses proposition*. It was stated in the second rule that the
" difference of the square of the sum of two numbers, and 4 times their rectangle,
" is equal to the square of their difference. Take then the squares of 23, it is
" 529, and 4 times the rectangle of the two sides, it is 480; their difference is
" 49, which is equal to the square of the difference of the sides, that is 7: then
" add 7 to 23, and subtract it from the same, and the halves, are 15 and 8 the
" two sides."

The next example is, $\diagup y$ required x, y, z, such that $x + y + z = 56$,
and $xyz = 4200$. " I suppose the diameter (*the hypothenuse*) unknown; take its
" square it is x^2: This is equal to the sum of the square of the two sides about
" the right angle, by the figure of the bride; and as 4200 is the product of the
" rectangle of the two sides multiplied by the hypothenuse, I divide 4200 by the
" unknown, the quotient $\dfrac{4200}{x}$ is the rectangle of the two sides. And it was
" stated that the excess of the square of the sum of the numbers above the sum
" of their squares is equal to twice the rectangle of the two numbers. The sum
" of the two sides is $56 - x$; I take its square, it is $x^2 - 112x + 3136$; and the
" sum of the squares of the two sides is x^2, for that is the square of the hypothenuse,
" which is the same. I take the difference of the two $- 112x + 3136$, and this
" is equal to twice the rectangle of the two sides, that is $\dfrac{8400}{x}$," &c.

The equation is reduced in the common way: the square in the quadratic,
which arises, being completed by adding the square of 14, which is half the co-
efficient of x. In this way the hypothenuse, and thence the other sides are
brought out.

---

* Meaning by the asses proposition the 20th of the first book of Euclid, which we are told was ridiculed by
the Epicureans as clear even to asses. These passages are only interpolations of the Persian translator.

END OF THE SECOND BOOK.

BOOK 3.

"EXPLAINING THAT MANY COLOURS MAY BE EQUAL TO EACH OTHER."

———◆———

"THE rule in this case is to subtract the unknown of one side from the un-
" known or cipher of the other side, and all the other colours and the numbers of
" the second side from the first side, from which the unknown was subtracted,
" and divide those colours by the unknown. If, as may happen, the denomi-
" nators are one quantity, perform the operation of the multiplicand; and if the
" denominators are different unknown quantities let them be unknown. Suppose
" the quantity of every one of these unknown the denominator, and put it below
" the colours of the dividend, and reduce the fractions and reject the denomina-
" tors; then the unknown will not remain on any side. After that subtract the
" black of one side from the other side, and subtract the rest of the colours and
" the numbers from the side from which the black was subtracted, and perform
" the same operations as were directed for the unknown, and the quantity of the
" black will be obtained; and in like manner the rest of the colours, and all
" the quantities of the multiplicand will be obtained. Then perform with it the opera-
" tion of the multiplicand; and the multiplicand and quotient will be obtained. The
" multiplicand will be the quantity of the dividend, and the quotient the quantity of
" the divisor. And if in the dividend of the operation of the multiplicand, two
" colours remain; as for example, black and blue, suppose the second in order,
" which is blue, the dividend, and suppose black a number, and add that to the
" augment, and perform the operation; and when the quantity of the two last
" colours is obtained, we shall known by the method which has been explained
" and illustrated in the examples, what are the quantities of the other colours
" which are below it. And when the quantity is known, reject the name of
" colour, and if the quantity of the colour is not obtained in whole numbers,
" again perform the operation of the multiplicand till it comes out whole; and by
" the quantity of the last colour we know the quantities of the other colours, so
" that the quantity of the unknown will be found. If then any one propose a
" question in which there are many things unknown, suppose them different

" colours. Accordingly, suppose the first unknown, and the second black, and
" the third blue, and the fourth yellow, and the fifth red, and the sixth green, and
" the seventh parti-coloured, and so on, giving whatever names you please to
" unknown quantities which you wish to discover. And if instead of these
" colours other names are supposed, such as letters, and the like it may be done.
" For what is required is to find out the unknown quantities, and the object in
" giving names is that you may distinguish the things required."

From the first question in this book arises the equation $5x + 8y + 7z + 90 =$
$7x + 9y + 6z + 62$. From this is derived $\dfrac{-y + z + 28}{2x}$ or $\dfrac{-y + z + 28}{2} = x$.

Now z is assumed $= 1$, and from $\dfrac{-y + 29}{2}$, the multiplicand and the quotient
are found by the rules of the fifth chapter of the introduction as follows: The
augment being greater than the divisor, the former is divided by the latter. The
quotient is retained, and the remainder is written instead of the augment; the
quotient is found $= 0$ and the multiplicand $= 1$. As the number of the quotients
arising from the division of the dividend by the divisor is in this case odd, and
as the dividend is negative; and each of these circumstances requiring the multiplicand to be subtracted from the divisor, and the quotient from the dividend,
the quantities remain as they were, viz. 0 and 1. Now adding 14, the quotient
of 29 divided by 2, to 0; the true quotient is 14 and the multiplicand $= 1$.
Therefore $x = 14$, and $y = 1$, and $z = 1$; and new values may be found by the
rules of the 5th chapter of the introduction.

The next question is the same as the third of the 1st book.

In the next we have the four quantities $5x + 2y + 8z + 7w$, and $3x + 7y +$
$2z + 1w$, and $6x + 4y + 1z + 2w$, and $8x + 1y + 3z + 1w$, all equal to each
other; and the values of x, y, z, and w are required. From the first and second
is found $2x = 5y - 6z - 6w$; from the second and third $3x = 3y + z - w$; and
from the third and fourth $2x = 3y - 2z + w$.

From the two first of these three equations $9y = 20z + 16w$, and from the two
last $3y = 8z - 5w$; whence $12z = 93w$; and dividing $\dfrac{31w + 0}{4} = z$; " and
" above, where the rule of the multiplicand was given, it was said that when the
" augment is cipher, the multiplicand will be cipher, and the quotient the quotient
" of the augment divided by the divisor; here then the multiplicand and quotient
" are both cipher." Then adding 31 for a new value of z, and 4 for a new

value of w, $31 = z$ and $4 = w$, and the other quantities are brought out in the usual manner.

The next example gives $5x + 7y + 9z + 3w = 100$, and $3x + 5y + 7z + 9w = 100$. From these comes $4y = -8z - 36w + 200$, and for the operation of the multiplicand $\dfrac{-8z - 36w + 200}{4} = y$. Suppose $w = 4$, then $-8z$ will be the dividend, and $+56$ the augment, and 4 the divisor. As 4 measures 56, 14 times without a remainder, the multiplicand will be $= 0$, and the quotient $= 14$: adding -8 to 14, and 4 to 0, $y = 6$ and $z = 4$. The other quantities are found in the same way as in the former examples. Another method, not materially different from the foregoing, is also prescribed for the solution of this question *.

A great part of the next example is not intelligible to me. What I can make out is this. To find x so that $\dfrac{x-5}{6} = y$, $\dfrac{x-4}{5} = z$, $\dfrac{x-3}{4} = v$, $\dfrac{x-2}{3} = w$, whole numbers. Taking values of x in these equations the following are found $6y = 5z - 1$, $5z = 4v - 1$, and $4v = 3w - 1$; from this last $w = 3$ and $v = 2$, but these numbers giving $\dfrac{7}{5}$ a fractional value of z, new values must be sought for w and v. Then after some part which I cannot understand, the author makes $w = 3 + 4u$, and says u is found $= 4$; then $w = 19$, $v = 2 + 3u$, $v = 14$. After more, which I cannot make out, he finds $\dfrac{4v-1}{5} = 11 = z$, by means of t which he adds to v and finds $= 15$. After more, which I can make nothing of, he finds $y = 9$ and $x = 59$.

The next example is, what three numbers are those which when the first is multiplied by 5 and divided by 20, the remainder and quotient will be equal; and when the second is multiplied by 7 and divided by 20, the remainder and the quotient will be equal, with an increase of 1, to the remainder and quotient of the first; and when the third is multiplied by 9 and divided by 20, in like manner, the remainder and quotient will be equal with an increase of 1 to the remainder and quotient of the second? The first remainder is called x, the second $x + 1$, and the third, $x + 2$, and these are also the quotients. Let the first number be y. By the

---

* From this place there is a great omission in my copy as far as the question $7x^2 + 8y^2 = \square$, and $7x^2 - 8y^2 - 1 = \square$, in the next book. Mr. Burrows's copy, however, being complete in this part, I shall proceed to supply the omission in mine from his.

question $\frac{5y}{20} = x + \frac{x}{20}$, whence $x = \frac{5y}{21}$. Let the second number be z, then $\frac{7z}{20} = x + 1 + \frac{x + 1}{20}$, whence $x = \frac{7z - 21}{21}$. Let the third number be v, then $\frac{9v}{20} = x + 2 + \frac{x + 2}{20}$, whence $x = \frac{9v - 42}{21}$. From the first and second values of x is found $\frac{7z - 21}{5} = y$, and from the second and third $\frac{9v - 21}{7} = z$. From this last is found by the operation of the multiplicand $z = 6$ and $v = 7$, and 9 is called the augment of z, and 7 the augment of v; as this value of z does not give y integer, other values must be sought. The augment of z is directed to be called w, and the value of w is to be sought; w is found $= 3$, and its augment 5; 33 is found by multiplying 3 by 9 and adding 6; at last the required numbers are found 42, 33, and 28. Most of this example after that part where z is found $= 6$, is unintelligible to me. It appears only that new values of z are found from $6 + 9w$, and that w and its values are found $w = 3$ and $3 + 5u$, and from $w = 3$ the numbers are found. I suppose the question is solved much in the same way as such questions are now commonly done.

The next question gives $\frac{x - 1}{2}, \frac{x - 2}{3}, \frac{x - 3}{5}, \frac{\frac{x - 1}{2} - 1}{2}, \frac{\frac{x - 2}{3} - 2}{3}, \frac{\frac{x - 3}{5} - 3}{5},$ to find x so that all these numbers shall be integers.

Let the number required be x; let the first quotient be $2y + 1$, this multiplied by the divisor 2 will produce for the dividend $4y + 2$, and 1 being added for the remainder $x = 4y + 3$. In like manner the second quotient being assumed $3z + 2$, $9z + 5 = 4y$; from this last, by the operation of the multiplicand, find $z = 3$ and $y = 8$, and the augment of z is $4v$, and that of y is $9v$; then $x = 4y + 3 = 3 + 4 \times (8 + 9v) = 35 + 36v$. As the value of y, 8 will not answer for x in the third condition, proceed thus: Let the third quotient be $5u + 3$. Multiply by 5 and add 3, $25u + 18 = x$, this is $= 35 + 36v$, hence $25u - 17 = 36v$; then by the operation of the multiplicand $u = 5$ and $v = 3$, $36v = 108$, $25 \times 5 - 17 + 35 = 143 = x$, and as $u = 5 + 36w$ and $v = 3 + 25w$, the augment of x is 900 because $25 \times 36 = 900$.

The next question is to find two numbers r and s such that $\frac{r - 1}{5}, \frac{s - 2}{6}, \frac{(s - r) - 2}{3}, \frac{(r + s) - 5}{9}$, and $\frac{rs - 6}{7}$ are integers. To find other numbers

K

besides 6 and 8. Let the first number be $5x + 1$, and the second $6x + 2$, the difference is $x + 1$. Divide by 3; suppose the quotient y and the remainder 2; then $x + 1 = 3y + 2$, and $x = 3y + 1$, and $5x + 1$ the first number $= 15y + 6$; and $6x + 2$ the second number $= 18y + 8$; their sum is $33y + 14$. Let $\dfrac{33y + 14}{9}$

$= z + \dfrac{5}{9}$; then $\dfrac{9z - 9}{33} = y = \dfrac{3z - 3}{11}$; from this is found $y = 3$ and $z = 12$, or $y = 0 + 3z$ and $z = 1 + 11w$; hence $5x + 1 = 45w + 6$, and $6x + 2 = 54w + 8$. "As the product of these taken according to the question involves w^2, and would "be a long work," suppose $45w + 6 = 51$ and let the second number be as it was; throw out 7 from both, the remainders are 2 and $5w + 1$; take their product, it is $10w + 2$. Divide by 7; suppose u the quotient and 6 the remainder; $\dfrac{10w + 2}{7} = u + \dfrac{6}{7}$, $10w + 2 = 7u + 6$, $\dfrac{7u + 4}{10} = w$; from this is found $w = 6$ or $6 + 7v$. The second number being $54w + 8$, it is $54 \times 6 + 8 = 332$, and its augment is $54 \times 7v = 378v$. As the first number is $45w + 6$, and was supposed $= 51$, its augment is $45 \times 7v$.

The next is, what number is that which being multiplied by 9 and 7 and the two products divided by 30, the sum of the two remainders and two quotients will be 26. "Suppose the number x, multiply it by 16, it is $16x$, for if I had "multiplied separately by 7 and 9, by the first figure of the second book, it would "also be $16x$". Let the quotient of $16x$ divided by 30 be y, $16x - 30y$ is the remainder, add the quotient y; $16x - 29y = 26$, and $\dfrac{29y + 26}{16} = x$. The augment being greater than the divisor, subtract 16 from 26, it is 10. By the operation of the multiplicand, the quotient is found 90 and the multiplicand 50. From 90 subtract the 29^s, and from 50 the 16^s; 3 and 2 remain. Take 3 from 29 and 2 from 16, 26 and 14 remain. As 16 was once rejected from the augment, add 1 to 26, $x = 27$, and the quotient is 14 and the remainder 12. No new values can be had in this case by the augment, for then the quotient and remainder would be greater than 27.

The next is, what number is that which multiplied by 3, 7, and 9, and the products divided by 30, and the remainders added together and again divided by 30, the remainder will be 11. Suppose the number x; let $19x$ be divided by 30, and let the quotient be y, then $19x - 30y = 11$. "If we had multiplied "separately, and divided each number by 30, the sum again divided by 30 would

" also have been equal to 11 ; but this would have been a long operation. The proof
" of the rule for such numbers is plain ; for example, if 8 be multiplied by 2, 3, and
" 4, it will be 16, 24, and 32, and dividing each by 15, there will remain 1, 9,
" and 2. The sum of these, that is 12, divide by 15 ; there remains 12. If 8 is
" multiplied by the sum of these that is 9, it will be 72 ; divide this by 15, 12
" remains." From $\frac{30y + 11}{19} = x$ by the operation of the multiplicand is found $x =$
29 + 30m, and $y = 18 + 19m$.

The next is, what number is that which being multiplied by 23, and divided
by 60, and again by 80, the sum of the remainders is 100 ? Let the number be x.
Suppose the first remainder 40, and the second 60, and let $\frac{23x}{60} = y + \frac{40}{60}$, then
$x = \frac{60y + 40}{23}$. Again, let $\frac{23x}{80} = z + \frac{60}{80}$, then $x = \frac{80z + 60}{23}$. Hence $80z + 20$
$= 60y$, from which are found $y = 3$ and $z = 2$; these values do not make x in-
teger. $y = 3 + 4m$, $z = 2 + 3m$. Let $y = 7$ and $z = 5$, then $x = 20$. By supposing the remainders 30 and 70, x will be $= 90$, and the question may be worked
without supposing the remainders given numbers, and by subjecting the quantities separately to the operation of the multiplicand.

In the next example y being the quotient of $\frac{5x}{13}$, $x + y = 30$. " Here there can
" be no multiplicand for no line (of quotients) is found, nor can it be brought
" out by interposition" (meaning quadratic equations). Proceed then by another
method and the question is solved by position ; the number is supposed 13, and
brought out truly $21\frac{2}{3}$; afterwards is added, " I say this too may be done by
" Algebra thus :" Call the number x,
$$\frac{5x}{13} + x = \frac{18x}{13} = 30, \quad 18x = 390, \quad x = 21\frac{2}{3}.$$
The next example is. It is said in ancient books that there were three people,
of whom the first had 6 dirhems, the second 8, and the third 100. They all
went trading and bought pawn leaves at one price, and sold them at one rate,
and to each person something remained. They then went to another place where
the price of each leaf was 5 dirhems ; they sold the remainder and the property of
the three was equal. At what price did they buy first, and at what rate did they
sell, and what were the remainders?

Let the number of leaves bought for 1 dirhem be x, and suppose the price they sold for to be a certain number. For example: Suppose 110 leaves sold for 1 dirhem, then the leaves of the first person were $6x$; let the quotient of $6x$, divided by 110 be y, which is the number of dirhems first had; $6x - 110y$ is the number of leaves remaining. Multiply by 5, $30x - 550y$ is their price; add the former result y, $30x - 549y$ is the amount of the first person's property. Then by four proportionals is found what the produce of $8x$ and $100x$ will be, that of $6x$ being y; the second person is found to have $y + \dfrac{y}{3}$ and the third $16y + \dfrac{2y}{3}$. After working as above, according to the terms of the question, the amount of the second person's property is found $\dfrac{120x - 2196y}{3}$ and in like manner the third person's $\dfrac{1500x - 27450y}{3}$. From $30x = 549y$, x is found $= 0$, and its augment 549, this is $= x$, and $y = 30$ *. It is added that unless a number is assumed the question cannot be solved without the greatest difficulty.

This book closes with some general remarks about the attention and acuteness requisite for solving questions like these.

---

* Some of these numbers are evidently brought out wrong, for x should be divisible by 5 and by 21. Taking 525 (instead of 549) for x, and putting a, b, c for the leaves sold at 110 per dirhem; we get $b = 1100 + a$, and $c = 51700 + a$; where a may be 110, or the multiples of 110 up to 770.

END OF THE THIRD BOOK.

BOOK 4.

" ON THE INTERPOSITION (توسيط) OF MANY COLOURS."

—⊙✳⊙—

" **A**ND that relates to making the squares of many colours equal to number.
" Its operation is thus: When two sides in the said condition are equal, in the
" manner that has been given above for the interposition of one colour, suppose
" a number and multiply or divide both sides by it, and add or subtract another
" number, so that one of the two sides may be a square. Then the other side
" must necessarily have a root, for the two sides are equal, and by the increase or
" decrease of equal quantities, equals result; then take the root of that which is
" easiest found. And if in the second there is the square of a colour and a
" number, suppose the square the multiplicand and the number the augment,
" and find the root by the operation of the square which was given above, and
" this certainly will be number. Make the first root of colours equal in these
" two, and know that you must equate so that the square, or the cube, or the
" square of the square, of the unknown may remain. And after the operation of
" the multiplication of the square, the less root is the quantity of the root of the
" square of the colour of that side which was worked upon; and the greater root
" is the root of all that side which was equal to the root of the first side. Equate
" then in these two sides. And if in the second side there is the unknown, or
" the square of the unknown, the operation of the multiplicand cannot be done.
" Then assuming the square of another colour perform the operation. Thus it is.
" If there is the unknown with numbers, or the unknown alone, whose root does
" not come out by the multiplication of the square, unless by assuming the square
" of another colour; when the root of this is obtained, equate in both and find
" the quantity of the unknown. The result of this is, that you must apply your
" mind with steadiness and sagacity, and perform the operation of multiplication
" of the square in any way that you can." Here follow a few lines of general
observations not worth translating.

Example. What number is that which being doubled, and 6 times its square added to it, will be a square?

Let the number be x, and let $2x + 6x^2 = y^2$. Multiply by 24, which is 6 multiplied by 4, and add 4; then divide by 4, it is $12x + 36x^2 + 1 = 6y^2 + 1$; $\sqrt{(12x + 36x^2 + 1)} = 6x + 1$. As the root of the other side $6y^2 + 1$ cannot be found, perform the operation of the multiplication of the square. Suppose the less root, or $y = 2$; then $6y^2 + 1 = 5^2$; $5 = 6x + 1$, $x = \frac{2}{3}$. By the rule of cross multiplication for new values $y = 2 \times 10 + 2 \times 10 = 40$ and $6x + 1 = 49$, whence $x = 8$.

The next is: What numbers are those two, the square of the sum of which, and the cube of their sum, is equal to twice the sum of their cubes?

Let the first number be $x - y$; and the second $x + y$, their sum will be $2x$: then $4x^2 + 8x^3 = 2((x - y)^3 + (x + y)^3) = 4x^3 + 12xy^2$; $4x + 4x^2 = 12y^2$; $4x^2 + 4x + 1 = 12y^2 + 1$; whence $2x + 1 = \sqrt{(12y^2 + 1)}$. Then by the multiplication of the square, making 2 the less root, 7 is the greater, $2x + 1 = 7$, $x=3$, $y=2$: $x-y=1$, $x+y=5$. By cross multiplication new values may be found.

The next is: What number is that which, when the square of its square is multiplied by 5, and 100 times its square subtracted from the product, the remainder is a square?

Let the number be x, and let $5x^4 - 100x^2 = y^2$; $5x^2 - 100 = \frac{y^2}{x^2} = \square$. Suppose 10 the less root, then $5 \times 10^2 - 100 = 400 = 20^2$; whence $y = 200$ and $x = 10$.

The next is: What are those two whole numbers whose difference is a square, and the sum of whose squares is a cube?

Let the two numbers be x and y; let $y - x = z^2$, then $x^2 = y^2 - 2yz + z^4$, and as $x^2 + y^2 = \boxed{}$, let $2y^2 - 2yz^2 + z^4 = z^6$. Then $2y^2 - 2yz^2 = z^6 - z^4$, and $4y^2 - 4yz^2 = 2z^6 - 2z^4$, and $4y^2 - 4yz^2 + z^4 = 2z^6 - z^4$; whence $2y - z^2 = \sqrt{(2z^6 - z^4)} = z^2 \sqrt{(2z^2 - 1)}$. Now by the multiplication of the square making 5 the less root, $2 \times 5^2 - 1 = 49$, and 7 is the greater root. Then $\sqrt{(2z^6 - z^4)} = 175 = 2y - z^2$; $2y - 25 = 175$, $y = 100$, $x = y - z^2 = 75$. Or if $z = 29$ new values of x and y will be found as above.

Here follows a Rule. "Know that when both sides are equal and the root of "one side is found, and on the other side there is a colour and its square, make "this side equal to the square of the next colour, that is to say not to x, and let

" its square be to that of y, and if y be its square, make it equal to the square of z,
" and multiply or divide both sides by a number, and add or subtract something
" so that the root of the side may be found. Here then I have found two roots;
" one the first, which is the root of the first of the first two sides: and the
" second, the root of the first of the second two sides, which is not equal to that
" root. Perform the operation of the multiplication of the square with that other
" side whose root is not found. Let the less root be equal to the first root, and
" the greater root be equal to the second root, and the quantity of these colours
" will be found."

Example. A person gave to a poor man in one day three units, and gave every
day with an increase of two. One day the poor man counted all the money,
and asked an accountant when he should receive three times the sum, at the rate
paid. Let the number of days passed when he counted his money be x, and the
number of days when the sum would be tripled y. First find the amount received
in the time x, thus: " By a rule in the Lilavati." $(x-1) \times 2 + 3 = 2x + 1 =$
the gift of the last day; $\dfrac{2x + 1 + 3}{2} = x + 2 =$ the gift of the middle day.
Multiply this by the number of days, $x^2 + 2x$ is the sum. In like manner the
sum for the time y is $y^2 + 2y$, which by the question is $= 3x^2 + 6x$; whence
$9x^2 + 18x + 9 = 3y^2 + 6y + 9$, and $3x + 3 = \sqrt{(3y^2 + 6y + 9)}$. Let $3y^2 +$
$6y + 9 = z^2$, then will be found $3y + 3 = \sqrt{(3z^2 - 18)}$. By the multiplication
of the square, making 9 the less root, $3z^2 - 18 = 15^2$, therefore $3y + 3 = 15$ and
$y = 4$; and because $3y^2 + 6y + 9 = z^2 = 81$, and $3x + 3 = 9$, $x = 2$. Thus,
on the first day, he got 3, and the second 5, and the sum is 8; and on the fourth
day he had 24, which is three times 8. In like manner, by making the less root
33, the greater root will be 57, and $y = 18$ and $x = 10$, and other values may be
found by assuming other numbers for the less root.

Then follows a Rule, which is so mutilated that I do not know how to translate
it. As far I can judge, its meaning appears to be this: If $ax^2 + by^2 = z^2$, the
quantities are to be found thus: Either find r such that $ar^2 + b = \square = p^2$, and then
x will be $= ry$ and $z = py$, or apply the rule given at the end of the 6th chapter
of the introduction for the case, when $a = \square$.

Required* x and y, such that $7x^2 + 8y^2 = \square$, and $7x^2 - 8y^2 + 1 = \square$

---

* At this place my copy comes in again.

$7x^2 + 8y^2$ is supposed $= z^2$. The operation of multiplication of the square is directed to be performed, $7x^2$ being the multiplicand, and $8y^2$ the augment: " I " suppose 2 the less root, and multiply its square which is four by 7 the multi- " plicand; it is 28; add 8, the square is 36; 6 then is found the greater root; " 6 black then is the quantity of blue, and 2 black the quantity of the unknown. ' Thus z is found $= 6y$ and $x = 2y$. ` The second condition is $7x^2 - 8y^2 + 1 = \square$, whence by substituting $2y$ for x; $28y^2 - 8y^2 + 1$, or $20y^2 + 1 = \square = w^2$. Now by the operation of multiplication of the square, supposing the less root, and $20 \times 2^2 + 1 = 81 = w^2$; whence $w = 9$. Therefore $x = 4$ and $y = 2$, supposing 36 the less root, x will be $= 72$ and $y = 36$.

In the next Example x and y are required such that $x^2 + y^3 = \square$, and $x + y = \square$. The multiplicand being a square let the augment be divided by y. Then by a rule of the 6th chapter of the introduction $\dfrac{y^2 - y}{2} = x^*$. Let $\dfrac{y^2 + y}{2} = w^2$, then $y^2 + y = 2w^2$. Multiply by 4 and add 1, $4y^2 + 4y + 1 = 8w^2 + 1$. The root of the first side of the equation is $2y + 1$. Find the root of the second side by the operation of multiplication of the square, supposing 6 the less root, 17 will be the greater; now $2y + 1 = 17$; whence $y = 8$ and $x = 28$ Other values of y and x are $49 = y$, and $1176 = x$.

Another method of solving this question is given. Supposing one of the numbers $2x^2$ and the other $7x^2$; the sum is $9x^2$, which is the square of $3x$. The square of the first added to the cube of the second, is $8x^6 + 49x^4$; let this be $= y^2$; divide by x^4, the quotient is $8x^2 + 49$. Perform the operation of multiplication of the square, supposing 2 the less root, $8 \times 4 + 49 = 81 = 9^2$. Therefore $x = 2$, and the first number $2x^2$ is $= 8$, and the second $7x^2$ is $= 28$; and supposing 7 the less root, 21 will be the greater root: then $x = 7$, and the first number will be 98 and the second 343.

Rule. " If a square is equal†, the root of which cannot be found‡, and in

---

* Viz. If A $= p^2$ (supposing $Ax^2 + B = y^2$), then $\dfrac{\frac{B}{n} + n}{2} = y$; and $\dfrac{\frac{B}{n} - n}{2}{p} = x$.

† Here seems to be an omission.

‡ If the number can be reduced to the form $(ax + my)^2 + ry^2$, it becomes rational by making $ax + my = \dfrac{r-1}{2}y$, for then $\left(\dfrac{r-1}{2}y\right)^2 + ry^2 = \left(\dfrac{r+1}{2}y\right)^2$. In Mr. Burrow's copy this rule begins, " If *there are two* " *sides,* the root," &c.

" which there are two squares of two colours, and the rectangle of those two
" colours; take the root of one square, and find from the second square a root, so
" that from the two squares that rectangle may be thrown out.

" For example: In the second side is 36 square of unknown, and 36 square of
" black, and 36 rectangle of unknown and black. Take the root of 36 square of
" unknown, 6 unknown; and from 36 square of black, take the root of 9 square
" of black, 3 black. When we take twice the rectangle of these two roots, the
" rectangle which is also 36 will be thrown out; and from the squares 27 square
" of black will remain. Divide whatever remains by the colour, of which this is
" the square; and from the number of the colour of the quotient, having sub-
" tracted one, halve the remainder, make what is obtained equal to that root
" which has been found. After dividing the second by the first, the quantity of
" the first colour will be obtained."

In the next example x and y are required such that $x^2 + y^2 + xy = \square$, and
$\sqrt{(x^2 + y^2 + xy)} \times (x + y) + 1 = \square$. " The first equation being multiplied by
" 36 gives $36x^2 + 36xy^2 + 36xy = 36z^2$. The root of one square and part of the
" second square, the rectangle having been thrown out, are found 6 unknown, and
" 3 black: there remains 27 square of black." Then applying the rule, x is

found $= \dfrac{5y}{3}$, whence $x^2 = \dfrac{25y^2}{9}$, and $x^2 + y^2 + xy = \dfrac{25y^2}{9} + y^2 + \dfrac{5y}{3}y = \dfrac{49y^2}{9}$

$= \left(\dfrac{7y}{3}\right)^2$, and $\sqrt{(x^2 + y^2 + xy)} \times (x + y) + 1 = \dfrac{7y}{3}\left(\dfrac{5y}{3} + y\right) + 1 = \dfrac{56y^2 + 9}{9}$; make

this $= w^2$, then $56y^2 + 9 = 9w^2$. Then root of $9w^2$ is $3w$, and by the operation
of multiplication of the square, making 6 the less root, 45 will be the greater root.
For $56 \times 36 + 9 = 2025 = 45^2$; therefore $y = 6$ and $x = 10$; or making 180
the less root, $y = 180$ and $x = 300$.

The next question is: Required x and y such that $\dfrac{xy + y}{2} = \boxed{}$, and $x^2 + y^2$
$= \square$, and $x + y + 2 = \square$, and $x - y + 2 = \square$, and $x^2 - y^2 + 8 = \square$, and
$\sqrt{\dfrac{x + xy}{2}} + \sqrt{(x^2 + y^2)} + \sqrt{(x + y + 2)} + \sqrt{(x - y + 2)} + \sqrt{(x^2 - y^2 + 8)}$
$= \square$. It is plain that 6 and 8 will answer the above conditions. Pass them
and find two others. It is required to find them by means of one unknown
quantity only. Suppose the first number $p^2 - 1$, and the second $2p$. Then
$\dfrac{xy + y}{2} = \dfrac{(p^2 - 1) \times 2p + 2p}{2} = \dfrac{2p^3 - 2p + 2p}{2} = p^3$. And $x^2 + y^2 = p^4 - $

$2p^2 + 1 + 4p^2 = p^4 + 2p^2 + 1 = (p + 1)^2$. And $x + y + 2 = (p^2 - 1) + 2p + 2$ $= p^2 + 2p + 1 = (p + 1)^2$. And $x^2 - y^2 + 2 = (p - 1)^2$. And $x^2 - y^2 + 8 =$ $p^4 - 2p^2 + 1 - 4p^2 + 8 = p^4 - 6p^2 + 9 = (p^2 - 3)^2$; and the sum of the roots is equal to $p + (p^2 + 1) + (p + 1) + (p - 1) + (p^2 - 3) = 2p^2 + 3p - 2$. As the root of this cannot be found, make it equal to 9^2: then $2p^2 + 3p = 9^2 + 2$. Multiply by 8 and add 9; $16p^2 + 24p + 9 = 89^2 + 25$. Find the root of the first side $\sqrt{(16p^2 + 24p + 9)} = 4p + 3$. For the root of the second side perform the operation of multiplication of the square. Suppose the less root 5, the greater root will be 15; for $8 \times 25 + 25 = 225 = 15^2$. Make the root of this equal to that of the first side $4p + 3 = 15$, whence $p = 3$. In this case $x = 8$ and $y = 6$; making the less root 30, the greater will be 85; p will then be $\dfrac{41}{2}$ and $x = \dfrac{1677}{4}$ and $y = 41$; or making the less root 175, the greater will be 495, $p = 123$, $x = 15128$, and $y = 266$. Or x may be supposed $= p^2 + 2p$, and $y = 2$. Or $x = p^2 - 2p$, and $y = 2p - 2$. Or $x = p^2 + 4p + 3$, and $y = 2p + 4$. And the numbers required may be brought out in an infinite number of ways besides the above.

Here follows an observation, that in calculation, correctness is the chief point ; that a wise and considerate person will easily remove the veil from the object ; but that where the help of acuteness is wanting, a very clear explication is necessary. " And so it is when there is such a question as this : What two numbers are those, " the sum or difference of which, or the sum or difference of the squares of " which, being increased or lessened by a certain number, called the augment, " will be a square. If examples of this sort are required to be solved by one " colour only, it is not every supposition that will solve them ; but first suppose " the root of the difference of the two numbers one unknown, and another " number with it either affirmative or negative. Divide the augment of the " difference of the two squares, by the augment of the sum of the numbers, and " add the root of the quotient to the root of the supposed difference abovemen- " tioned ; it will be the root of the two numbers. Take then every one, the " square of the root of the difference of the numbers, and the square of the root " of the sum of the numbers, and write them separately. Afterwards, by the way " of opposition add and subtract, the augment of the difference, and the sum of " the two numbers aforementioned, as is in the example, to and from the squares of " the two, which, by the question, were increased or diminished. The result of " the addition and subtraction will be known, and from that the two numbers

" may be found in this manner, viz. by the rule $\dfrac{(x+y)+(x-y)}{2} = x$, and

" $\dfrac{(x+y)-(x-y)}{2} = y$."

The next is to find x and y such that $x + y + 3 = \square$, and $x - y + 3 = \square$, and $x^2 + y^2 - 4 = \square$, and $x^2 - y^2 + 12 = \square$, and $\dfrac{xy}{2} + y = \boxed{}$, and the sum of the roots $+ 2 = \square$. Exclude 6 and 7, which it is plain will answer. $\sqrt{}(x - y)$ is supposed $= p - 1$, then x is made equal to $p^2 - 2$ and $y = 2p$, wherefore $x + y + 3 = (p^2 - 2) + 2p + 3 = (p + 1)^2$, and $x - y + 3 = (p^2 - 2) - 2p + 3 = (p - 1)^2$, and $x^2 + y^2 - 4 = (p^4 - 4p^2 + 4) + 4p^2 - 4 = (p^2)^2$, and $x^2 - y^2 + 12 = (p^4 - 4p^2 + 4) - 4p^2 + 12 = (p^2 - 4)^2$, and $\dfrac{xy}{2} + y = \dfrac{(p^2 - 2)\,2p}{2} + 2p = p^3$, and the sum of the roots $+ 2 = (p + 1) + (p - 1) + p^2 + (p^2 - 4) + p + 2 = 2p^2 + 3p - 2$; make this $= q^2$; $2p^2 + 3p - 2 = q^2$, and $2p^2 + 3p = q^2 + 2$. Multiply by 8 and add 9. $16p^2 + 24p + 9 = 8q^2 + 25$. The root of the first side is $4p + 3$. Find the root of the second side by the operation of multiplication of the square ; making the less root 175, the greater root will be 495. Therefore $4p + 3 = 495$, and $p = 123$, and $x = 15127$, and $y = 246$.

The next is : Required x and y such that $x^2 - y^2 + 1 = \square$, and $x^2 + y^2 + 1 = \square$. Let $x^2 = 5p^2 - 1$ and $y^2 = 4p^2$, $x^2 - y^2 + 1 = p^2$ and $x^2 + y^2 + 1 = (3p)^2$. The root of $4p^2$ is $2p$. Find the root of $5p^2 - 1$ by the operation of multiplication of the square. Supposing the less root 1, the greater will be 2. Supposing 17 the less root, the greater will be 38. Or if $x^2 + y^2 - 1 = \square$, and $x^2 - y^2 - 1 = \square$, let $x^2 = 5p^2 + 1$ and $y^2 = 4p^2$; and so on as in the first case.

Rule. " When the root of one side is found, and on the second side there is
" a colour, whether with or without a number, equate that side with the square
" of the colour which is after it and one unit. And bring out the quantity of
" the colour of the second side which is first in the equation ; and bring out what
" is required in the proper manner."

Example. To find x and y such that $3x + 1 = \square$, and $5x + 1 = \square$. Let $3x + 1 = (3z + 1)^2$, then $x = 3z^2 + 2z$, let $5(3z^2 + 2z) + 1 = w^2$, whence $15z^2 + 10z = w^2 - 1$; multiply by 15 and add 25 ; $225z^2 + 150z + 25 = 15w^2 + 10$. The root of the first side is $15z + 5$. Find the root of the second side by the operation of multiplication of the square ; making the less root 9, the greater will be 35 ; $15z + 5 = 35$; therefore $z = 2$ and $x = 16$. By another way : Let $x = \dfrac{z^2 - 1}{3}$; multiply by 5 and add 1, $\dfrac{5z^2 - 2}{3}$, make this $= w^2$; $5z^2 = 3w^2 + 2$. Multiply by 5 ; $25z^2 = 15w^2 + 10$; the root of the first side is $5z$. Find the the root of the second as before, making the less root 9, the greater will be 35 ; whence z and x. In the above example other values of x are mentioned besides those which I have taken notice of.

The next example is : Required x such that $3x + 1 = \boxdot$, and $3(3x + 1)^{\frac{2}{3}}$ $+ 1 = \square$. Let $3x + 1 = y^3$; then $3x = y^3 - 1$, and $x = \dfrac{y^3 - 1}{3}$; multiply this by 3 and add 1, the result is y^3, the cube cube root of which is y. Let $3y^2 + 1 = z^2$; making the less root 4, the greater will be 7, whence $x = 21$.

The next is : To find x and y such that $2(x^2 - y^2) + 3 = \square$, and $3(x^2 - y^2) + 3 = \square$. " Know that in bringing out what is required, you must sometimes suppose the " colour in that number which the question involves, and sometimes begin from " the middle, and sometimes from the end, whichever is easiest. Here then " suppose the difference of the squares unknown," &c.

Let $x^2 - y^2 = p$; make $2p + 3 = q^2$; then $\dfrac{q^2 - 3}{2} = p$; multiply this by 3, and add 3, it is $\dfrac{3q^2 - 3}{2}$; let this be $= r^2$; therefore $3q^2 - 3 = 2r^2$; multiply by 3 and transpose ; $9q^2 = 6r^2 + 9$; the root of the first side is $3q$. Find that of the second side by the operation of multiplication of the square. Making the less root 6, the greater will be 15. Or making the less 60, the greater will be 147. If $3q = 15$, $q = 5$; if $3q = 147$, $q = 49$. In the first case $p = 11$, and in the second $p = 1199$. Suppose $x - y = 1$, $x^2 - y^2$ being $= 11$, $\dfrac{x^2 - y^2}{x - y} = x + y = \dfrac{11}{1} = 11$; and $x + y$ and $x - y$ being given, x and y may be found. In the first case $x = 6$ and $y = 5$. In the second $x = 600$ and $y = 599$.

Rule " If the square of a colour is divided by a number and the quotient is " a colour. If after the reduction of the equation its root is not found, make it " equal to the square of a colour, that the quantity of the black may come " out."

The next example which concludes this book is: Required x such that $\dfrac{x^2 - 4}{7}$ $=$ a whole number. Make $\dfrac{x^2 - 4}{7} = y$, then $x^2 = 7y + 4$: the root of the first side is x; that of the second side cannot be found. "Then by the above rule" let $7z + 2 = \sqrt{(7y + 4)}$; $49z^2 + 28z + 4 = 7y + 4$; whence $7z^2 + 4z = y$. "As the quantity the of black is 7 square of the blue, and 4 blue; and as 7 blue and "2 units were supposed equal to a root which is equal to the unknown, I make "it equal to the unknown. This same is the quantity of the unknown. I sup- "pose the quantity of the blue a certain number," &c. As $7z + 2 = x$ If $z = 0$, $x = 2$. If $z = 1$, $x = 9$. If $z = 2$, $x = 16$. Other values of x may be found in the same manner.

"After* equating that the two sides may come out, multiply the first side by "a number and take its root, and keeping the second side as it was, multiply the "number of the second side by the number which the first was multiplied by, "and make it equal to the square of a colour."

Example. What number is that whose square being multiplied by 5, and 3 added, and divided by 16, nothing remains? Let the number be x. Let $\dfrac{5x^2 + 3}{16}$ $= y$, a whole number; then $5x^2 = 16y - 3$, $5x^2 \times 5 = 25x^2$, $\sqrt{(25x^2)} = 5x$. Then there seems to be assumed $3 \times 5 = z^2 - 1$, and afterwards from $\dfrac{8z + 1}{5} = x$, the question is prepared for solution.

The next rule is: "If the cube of a colour is divided by a number, and the "quotient is a colour, make it equal to the cube of a colour. The way to find "that, is this: Assume the cube of a number and divide it by the divisor; there "should be no remainder; and add the number with it again and again to the "divisor, or subtract it from it; or let the cube be a cube of a number, which "join with it; or again multiply that number by the fixed number 3, and the "result multiply into the quotient, and divide it by the dividend; also there

---

* In Mr. Burrow's copy the fourth book ends with two rules and two examples, which, as far as I can make them out, are as above.

" will be no remainder. If a number can be found with these conditions equate
" with its cube."

Example.   What number is that from whose cube 6 being taken and the remainder divided by 5 nothing remains? Let the number be x and $\dfrac{x^3 - 6}{5} = y$
a whole number.   Hence $x^3 = 5y + 6$.   Then the cube root of this which is $= x$
is assumed $= 5z + 1$, and y is found $= 25z^3 + 15z^2 + 3z - 1$.

END OF THE FOURTH BOOK.

BOOK 5.

" ON THE EQUATION OF RECTANGLES."

—◆—

" $\mathbf{A}$ND that relates to the method of solving questions which involve the
" rectangles of colours. Know that when the question is of one number multi-
" plied by another, if the two numbers are supposed colours, it necessarily comes
" under rectangle of colours. The solution of that being very intricate and
" exceedingly difficult, if one number is required suppose it unknown; and if
" two or three, suppose one unknown and the others certain numbers, such that
" when they are multiplied together according to the question, no colour will be
" obtained except the unknown, and it will not come under rectangle of colours.
" And besides multiplication, if the increase or diminution of a number is re-
" quired, perform the operation according to the question, then it will be exactly
" a question of the same sort as those in the first book, which treats of the
" equality of unknown and number. By the rules which were given there, what
" is required will be found."

The first question is to find x and y such that $4x + 3y + 2 = xy$. Supposing
$y = 5$, then $4x + 17 = 5x$, wherefore $x = 17$ and $y = 5$. Supposing $y = 6$,
then $x = 10$. In like manner any number whatever being put for y the value of
x will be found.

The next is to find w, x, y, z, such that $(w+x+y+z) 20 = wxyz$. Suppose
the first w, the second 5, the third 4, and the fourth 2; then $20w+220=40w$,
and $w = 11$. Other values of w, x, y, z, are taken notice of.

The next is to find x and y in integers such that $\sqrt{(x + y + xy + x^2 + y^2)} +$
$x + y = 23$, or $= 53$. In the first case, suppose the first number x, and the
second 2, then $\sqrt{(x^2 + 3x + 6)} + x + 2 = 23$, and $\sqrt{(x^2 + 3x + 6)} = 21 - x$,
and $x^2 + 3x + 6 = x^2 - 42x + 441$; whence x will be found $= \dfrac{29}{3}$; this not
being an integer, let the operation be repeated. Suppose $y = 3$ then x will be

found $= \dfrac{97}{11}$; this too being a fraction, suppose $y = 5$; then x will be $= 7$. In the second case a number is put for y, and a fractional value of x is found. " And if we suppose the second number 14, the quantity of the unknown will " be 17, and this is contrary; for if the second number is supposed 17, the quan- " tity of the unknown will be 11 ; and if one is supposed a colour and the other a " certain number, it is probable that the unknown will be brought out a fraction; " and if a whole number is required, it may be found by much search, And if " both are supposed colours, and the question solved by this rule, a whole " number will easily be found,"

Rule*. " When two sides are equal, the method of equating them is thus : " subtract the rectangle of one side from the other side, and besides that what- " ever is on the second side is to be subtracted from the first ; then let both sides " be divided by the rectangle ; and on the side where there are colours let those " colours be multiplied together. And let a number be supposed, and let the " numbers which are on that side be added to it ; and let the result be divided " by the supposed number ; and let the quotient and the number of the divisor " be separately increased or lessened by the number of the colours which were " before multiplication, whichever may be possible. Wherever the unknown is " added or subtracted there will be the quantity of the black ; and wherever the " black is added or subtracted there will be the quantity of the unknown. And " in like manner if there is another number, and if both addition and subtraction " are possible, let both be done, and two different numbers will be found. Also " if the number of the colours is greater, and cannot be subtracted, subtract the " quotient and the number of the divisor from the colour if possible, what was " required will be obtained."

Example. x and y are required such that $4x + 3y + 2 = xy$. Multiply 3 by

---

* This rule is very ill expressed ; it must mean—The equation being reduced to $ax + by + c = xy$, $a +$ $\dfrac{ab+c}{p}$ will be $= y$ and $b + p = x$. Because $ax+by+c=xy$, $c = xy - ax - by$, add ab to both sides, then $ab+c$ $= xy - ax - by + ab = (x - b)(y - a)$: and making $p = x - b$, $y - a$ will be $= \dfrac{ab+c}{p}$. Therefore $x = b +$ p, and $y = a + \dfrac{ab+c}{p}$. More formulæ may be had by resolving $ab + c$ into different factors.

4, and add 2 to the product $3\times4+2=14$. Suppose 1. Divide 14 by 1. Add 4 to the number 1, and add 3 to the quotient $4+1=5=y$, and $3 + \dfrac{14}{1} = 17 = x$.

Or $4 + \dfrac{14}{1} = 18 = y$, and $3 + 1 = 4x$. " And no other case is possible." Dividing by 2, the quotient will be 7; $y = 11$ and $x = 5$. And by another method y will be found $= 6$ and $x = 10$*.

To find x and y so that $10x + 14y - 58 = 2xy$. After reducing the equation to $5x + 7y - 29 = xy$. By the rule above given, assume divisors of $5\times7-29$;— 1 being the divisor, 6 is the quotient.

$$5 + 1 = 6 = y \text{ and } 7 + 6 = 13 = x,$$
$$\text{or } 5 + 6 = 11 = y \qquad 7 + 1 = 8 = x,$$
$$\text{or } 5 - 1 = 4 = y \qquad 7 - 1 = 6 = x.$$

2 being the divisor, 3 is the quotient.

$$y = 8, \qquad y = 7, \qquad y = 3,$$
$$x = 9, \qquad x = 10, \qquad x = 4;$$

and no others are possible. 3 being the divisor 2 is the quotient, and the quantities are as above. It is added that these two examples may be proved by geometrical figures as well as numbers.

---

* In Mr. Burrow's copy there is another example which is wanting in mine. It is as above.

THE END.

Mr. Davis's Notes.

I HERE put together all I have been able to make out of Mr. Davis's notes of the Bija Ganita. What I have extracted literally is marked by inverted commas; the rest is either abstract, or my own remarks or explanations. I have preserved the divisions of the Persian translation for the convenience of arrangement and for easy reference. Mr. Davis's letter to me, authenticating these notes, is annexed.

Chapter 1st of Introduction.

THE manner in which the negative sign is expressed, is illustrated in the notes by the addition and subtraction of simple quantities, thus: " Addition.—When " both affirmative or both negative, &c. When contrary signs, the difference " is the sum.

$$
\begin{array}{cccc}
\dot{3} & 3 & 3 & \dot{3} \\
4 & 4 & \dot{4} & 4 \\
\hline
\dot{7} & 7 & \dot{1} & 1 \\
\hline
\end{array}
$$

" Subtraction.

$$
\begin{array}{cccc}
3 & \dot{3} & 3 & \dot{3} \\
2 & \dot{2} & \dot{2} & 2 \\
\hline
1 & \dot{1} & 5 & \dot{5} \\
\hline
\end{array}
$$

" Multiplication.

" When both are affirmative or both negative the product is affirmative.

$$2 \times 3 = 6, \quad \dot{2} \times \dot{3} = 6, \quad 2 \times \dot{3} = \dot{6}, \quad \dot{2} \times 3 = \dot{6}.$$

" Why is the product of two affirmative or two negative quantities always
" affirmative? The first is evident. With regard to the second it may be ex-
" plained thus: Whether one quantity be multiplied by the other entire, or in
" parts, the product will always be the same, thus:

$$\text{`` } 135 \times 12 = 1620$$
$$\text{`` } 135 \times 8 = 1080$$
$$\text{`` } 135 \times 4 = 540$$
$$1620$$

" Then, let 135 be $\times$ by $\dot{4}$, but $12 - \dot{4} = 16$ and $135 \times \dot{4} = 540$; 135×16
" $= 2160$, and $540 + 2160 = 2700$, which is absurd: but $540 + 2160 = 1620$."

Mr. Davis remarks to me that there are here evidently some errors and some
omissions, and he thinks that the meaning of the last part of the passage must
have been to this effect: 12 may be composed of 16 added to $\dot{4}$. Let 135 be
multiplied by 12, so composed

$$135 \times 16 = 2160 \qquad\qquad 135 \times 16 = 2160$$
$$135 \times 4 = 540 \qquad\qquad 15 \times \dot{4} = 540$$
$$135 \times 12 = 2700 \text{ This is absurd: but } 135 \times 12 = 1620 \text{ which is}$$

right. Thus too $\dot{4}$ may be taken as formed by $12 + \dot{16} = \dot{4}$, and if

$$135 \times \dot{16} = 2\dot{1}60 \qquad\qquad 135 \times \dot{16} = 2160$$
$$1 5 \times 12 = 1\dot{6}20 \qquad\qquad 135 \times 12 = 16\dot{2}0$$

$135 \times 4 = 3780$ which is absurd: but $13\dot{5} \times \dot{4} = 540$ which is right.

Perhaps something like the following might have been intended :

$-135 \times -12 = 1620$ either $+$ or $-$; $\left. \begin{array}{l} -135 \times -8 \\ -135 \times -4 \end{array} \right\} = 1620$ either $+$ or $-$; now $4 - 12 =$

-8; and $8 - 12 = -4$; therefore the sum of $\left. \begin{array}{l} -135 \times (4-12) \\ -135 \times (8-12) \end{array} \right\}$ must be $= -135 \times -12$,

$$-135 \times (4-12) = -\ 540 + \text{ or } -1620$$
$$-135 \times (8-12) = -1080 + \text{ or } -1620$$

product -1620 $-3240 = -4860$ if $-\times-$ gives $-$; but $-1620 + 3240 = +1620$ if $-\times-$ gives $+$; therefore $-135 \times -12 = +1620.$

Chapter 3.

" OF QUANTITIES UNKNOWN, BUT EXPRESSED BY LETTERS."

" Jabut tabut . . . . . . 1st . . या

" Kaluk . . . . 2d . . का

" Neeluk . . . . 3d . . नी

" Peet . . . . 4th . पी

" Loheet . . . . 5th . लो

&c.

" Commentary adds Hurretaka . . . 1

" Chitraka 2

&c.

" These are styled abekt or unknown.

" These may be added to themselves, subtracted, &c. but cannot be added " to, &c. known quantities in the manner explained, or to unlike quantities of " any kind. The square of या cannot be added to या, but the addition may be " expressed thus या 1 add to या या ; the reason is, because to add 5 signs " to 2 degrees we cannot say 5 added to 2 is equal to seven, for this would be " absurd, we therefore write the sum $5^s\ 2^o$. But when the unknown quantity is " discovered it may then be added to the known, into one simple quantity.

" The unknown quantities are usually written first, and the highest powers of
" them before the lower..

" या व२ | या३ | 3. This is $2x^2 + 3x + 3$.

Also

" The multiplication of unknown quantities.

" To multiply या२ | into २ | we have या४.

या by या gives its square or या व, and this multiplied by या, gives या घ, &c.

Also this example of multiplication.

$$
\begin{array}{cc|c|c||c|c}
\text{" या३} & \text{या५} & \text{लो̇ı} & \text{या व१५} & \text{या३} \\
\text{" लो२} & \text{या५} & \text{लो̇ı} & \text{या१०} & \text{लो२} \\
\hline
& & & \text{या व१५} & \text{या७} & \text{लो२}
\end{array}
$$

which is the product of $(5x - 1) \times (3x + 2)$.

Chapter 4.

" OF THE CARNI OR SURD QUANTITIES."

" Example of two numbers, 2 and 8.

" $2 + 8 = 10$, the mahti carni.

" $2 \times 8 = 16$; its root is 4, and $4 \times 2 = 8$ the laghoo carni.

" The mahti carni 10

" Laghoo carni added . . 8
_____

18, the sum of these carnis. This 18 is the square of
" the sum of their roots."

And there is another example with the numbers 4 and 9, and the following

theorem, " 2)8(4, its root is 2, $\overset{2}{+} 1 \overset{2}{-} 1$

$$\frac{}{3 \quad\quad 1}$$

"3 × 3 = 9   9 × 2 = 18 sum.

"1 × 1 = 1   1 × 2 = 2 difference."

Also this : " The carni 18 is found ; its root is the sum of the roots of the two
" given numbers; but if there be two roots there must be two squares, the
" difference is the square of the difference between these squares."

And the following examples in multiplication :   " To multiply the square roots
" of 2, 8, and 3, by the square root of 3 and the integral number 5.

" These are surds, therefore take the square of the sum of the square roots of 2
and 8, and multiply by the square of 5.

" Square of sum of square roots of 2 and 8 is 18.

18	25, 3	450, 54	
3	25, 3	75, 9	root of 9 is 3 roop.

Sqrs.   Sqrs.   Sqrs.

450   54   75 — roop 3.

" Example second.

			Roop.	Carni.	Carni.
" Multipliers			5,	3,	12.

" 5	25	25, 3	625, 75	25 roop	25 roop	675
" 3⟩ 12⟨	27	25, 3	675, 81	9 roop	9 roop	75
				16	300	

" The product therefore is 16 roop, 300 carni."

The square of a negative quantity being made negative is here taken notice of
as in the Persian Translation : In division the following rule is mentioned.

" The carni divisor : reverse of each term, its sign, and multiply both divisor
" and dividend."

Carni which here means surd, means also the hypothenuse of a right-angled
triangle.

Chapter 5.

———◦✱◦———

" WHAT is that number by which when 221 is multiplied and 65 added to the
" product, and that product divided by 195, nothing will remain.

" The dividend *bhady*, divisor *hur* or *bhujuk*, the number added or subtracted
" is called *chepuk*. The bhady is here 221, the bhujuk 195; when divided the
" quotient is 1, this is disregarded; the *seke* or remainder is 26, by which 195
" divided the quotient is 7 disregarded, the remainder is 13, by which divide
" 221, the quotient is 17, the remainder is 0. The quotient 17 is the true or
" *dirl-bhady*.

" Then 95 divided by 13, the quotient is 15; the remainder 0. This quotient
" is named *dirl-bhujuk.*

" Then divide 65 by 13, the quotient is 5; the remainder 0; the quotient is
" the *dirl-chepuk.*

" They are now reduced to the smallest numbers.
   " 17 dirl-bhady.
   " 15 dirl-bhujuk.
   " 5 dirl chepuk.

The quotients are found and arranged as in the rule with 5 and 0 below,
  " 1
  ———
thus: 7
  ———
  5
  ———

  0 this is called *bullee*; the cipher is called *unte* or the latter; the next (5)
" is called *upantea.* Multiply this by its next number (7) and add the next below
" 5, this being 0, the product will be 35. Multiply this by the uppermost number
" (1) and add the next below (5) the amount is 40."

Then 40 and 35 are directed to be divided by the dirl-bhady and bhujuk.

17) 40 (2

34
———

6 this is called *lubd*.

15) 35 (2

30
———

5 this is called *goonuk*, and it is the number sought.

$$\frac{221 \times 5 + 65}{195} = 6,$$ and directions are given for finding new values of x and y,

(supposing $\dfrac{ax + c}{b} = y$) by adding a (in its reduced state) and its multiples to

the value of y ; and b and its multiples to the value of x.

The next question in the notes is also the same as that in the Persian.

" Bhady 100, bhujuk 63, and chepuk 90.

" OPERATION.

" These numbers cannot be all reduced to lower proportionals.

" 100 divided by 63, the quotient is 1, the remainder 37 ; by this remainder
" divide 63, the quotient is 1, the remainder is 26 ; by this divide 37, the quotient
" is 1, the remainder 11. Divide again ; quotient 2, remainder 4. Divide
" again ; quotient 2, remainder 3. Divide again ; quotient 1, remainder 1 ; this re-

1
———

" mainder 1 is disregarded. The several quotients write down thus : 1
———

1
———

2
———

2
———

1
———

90 the chepuk.
———

0

" Multiply and add from the bottom as in the former example, $90 \times 1 + 0 = 90$,
" $90 \times 2 + 90 = 270$, $270 \times 2 + 90 = 630$, $630 \times 1 + 270 = 900$, $900 \times 1 + 630 = 1530$,
" $1530 \times 1 + 900 = 2430$.

" The two last are the numbers sought; then

" 100)2430(24 this is disregarded.

$$200$$

$$430$$
$$400$$

30 Seke or remainder is the lubd.

" 63)1534(24
$$126$$

$$270$$
$$252$$

18 this is the goonuk."

$$\frac{100 \times 18 + 90}{63} = 30.$$

The method of reducing the bhady and chepuk is noticed, and the values of
$x = 171$ and $y = 27$, being first found the true values are found, thus:

$$63)171(2 \quad \text{and} \quad 10)27(2$$
$$126 \qquad\qquad 20$$

$$45 \qquad\qquad 7$$

$63 - 45 = x$ and $(10 - 7) \times 10 = y.$

The several methods of proceeding: first, by reducing the bhady and chepuk;
second, by reducing the bhujuk and chepuk; third, by reducing the bhady and
chepuk; and then the reduced chepuk and the bhujuk are also mentioned.

The following explanation of these reductions is given:

" The bhady 27, bhujuk 15;

" these are divided each by 3 . . . . 9 and 5.

" Write 27 in two divisions . . . . . 9 and 18

" these again divided by 3 . . . . . 3 and 6

" these two add $3 + 6 = 9$; thus the parts added, how many so ever are, always
" equal to the whole, thus therefore they are reduced to save trouble, and there-
" fore all these numbers are so reduced; but the goonuk is as yet unknown. Let

N

" it be supposed to be 5, by which multiply the parts of the bhady 9 and 18;
" 9 × 5 = 45, 18 × 5 = 90, which added are 135, and the bhady 27 × 5 =
" the same 135; this divided in two parts, 60 and 75. and added again, are 135.
" The lowest terms of 27 and 15 above, are 9 and 5; the common measure 3,
" multiplied by 5, 3 × 5 = 15 and 9 × 15 = 135.

" Thus too the chepuk must be reduced, and when they are all reduced to the
" lowest, the lubd and goonuk will be true; and if their numbers are not reduced
" to their lowest terms, the work will be the greater."

The principle on which the chepuk is reduced is explained thus:

" OF THE CHEPUK."

" The bhady 221, bhujuk 195, chepuk 65; the goonuk was found 5, lubd 6.

$$221 × 5 = 1105$$
$$195)1105(5 \text{ lubd.}$$
$$975$$
$$\overline{}$$

        130 seke, which deduct from the bhujuk 195 − 130 = 65 equal
" to the chepuk, which divide by the bhujuk 195)195(1. The lubd is 5, to
" which add 1; 6 = the original lubd."

In another example the bhady = 60, bhujuk = 13, and chepuk = 16 or − 16.
By the bullee are found the numbers 80 and 368; then 368−60×6=8 the lubd,
and 80 − 13 × 6 = 2 the goonuk; 60 − 8 = 52 the lubd corrected, and
13 − 2 = 11 the goonuk corrected. $\dfrac{60×11+16}{13}=52$, and $\dfrac{60×2-16}{13}=8$.

" Note in the text: The product by the two uppermost terms of the bullee,
" when divided by the bhady and bhujuk respectively, have hitherto always
" quoted the same number, as in the last example 6 the quotient, and the like
" also in the foregoing examples, but when it happens otherwise, as in the fol-
" lowing: When the bhady is 5, the bhujuk 3, the chepuk 23 affirmative or
" negative, what will be found the goonuk?

$$3)5(1$$
$$3$$
$$\overline{}$$
$$2)3(1$$
$$2$$
$$\overline{}$$

1 seke disregarded

$$\left.\begin{array}{c} 1 \\ \overline{} \\ 1 \\ \overline{} \\ 23 \\ \overline{} \\ 0 \end{array}\right\} \text{Bullee.}$$

23 × 1 = 23, + 0 = 23   5) 46 (9   3) 23 (7
23 × 1 = 23, + 23 = 46   45   21
                                             1   2 goonuk

" The two quotients being different numbers they must be taken the same;
" thus instead of 9, take the quotient 7.

$$5)46(7$$
$$35$$
$$\overline{11}$$

" therefore the goon is 2, the lubd 11. $\dfrac{5 \times 2 + 23}{3} = 11.$

" Next, when the chepuk is negative, or to be deducted, the rule directs to
" subtract the lubd from the bhady, but here it cannot be done: the rule is
" reversed, thus $11 - 5 = 6$, which is the lubd for the negative chepuk; next for
" the goon of the rhin chepuk $3 - 2 = 1$; therefore the goon and lubd for the
" rhin chepuk are 1 and 2; $5 \times 1 = 5$; but from this the rhin chepuk cannot be
" taken; therefore take it from the chepuk $23 - 5 = 18.$

$$" 3)18(6 \text{ the lubd.}"$$
$$18$$
$$\overline{0}$$

Other cases are mentioned for the negative chepuk, and for the chepuk reduced, and for new values of the goon and lubd.

The examples $\dfrac{5x + 0}{13}$ and $\dfrac{5x + 65}{13}$, which are in the Persian translation, are
also stated here, but no abstract of the work is given, only the lubd is said to be
5 and the goonuk 0, which applies to the last of the two only.

" The seke in bekullas is termed sood, meaning that it is the chepuk; the
" bhady, let it be 60. The coodin or urgun is the bhujuk, from which the lubd
" will be found in bekullas, and the goon will be the seke of the cullas, which
" must be taken as the chepuk; making the bhady again 60, the bhujuk will be
" the urgun, the lubd of this will be in cullas, the seke is the seke of the ansas,
" which seke must be taken as the chepuk; the bhady being taken 30, the
" bhujuk is still the urgun, the lubd is in ansas, the seke is the seke of the signs,

" which seke take as the chepuk ; making the bhady 12, the bhujuk will be still
" the coodin, the lubd here will be signs, the seke is the seke of bhaganas,
" revolutions, which seke must be taken as the chepuk ; the lubd will here be
" in bhaganas, the seke the urgun."

Example. " Let the calp coodin or urgun be 19, the bhaganas 9, the
" urgun 13."

" Then by proportion if 19 gives 9, what will 13 give ?" This is found to be
6 rev. 1 sign, 26°, 50′, 31″, with a fraction of 11 ; then from $\dfrac{60x - 11}{19} = y$,
x and y are found $x = 10$, $y = 31$; then from $\dfrac{60x' - 10}{19} = y'$, $y' = 50$ and $x' = 16$,
from $\dfrac{30x'' - 16}{19} = y''$, $y'' = 26$, $x'' = 17$, from $\dfrac{12x''' - 17}{19} = y'''$, $y''' = 1$, $x''' = 3$
from $\dfrac{9x'''' - 3}{19} = y''''$, $y'''' = 6$ and $x'''' = 13$, which is the urgun.

In another Example. Seke bekullas $= 11''$, bhaganas $= 49$, calp coodin or
urgun $= 149$, Jeist urgun $= 97$. The quantity is found by the rule to be $=$
23 rev. 10 signs, 18°, 23′, 31″, the remainder 11.

" The addy month 1, is the bhady ; the coodin 195, the bhujuk ; the seke of
" the addy month 95, is the chepuk.

" 195)1(0   0

1 seke disregarded   95 bullee.

                                             0

" $0 \times 95 + 0 = 0$ Raas $\dfrac{0}{95}$

" 1)0(0   195)95(0

0 lubd   95 goonuk.

" $95 \times 1 - 95 = 0$ 195)0(0

                                             0

" The che tits 26 is the bhady; coodin 225, is the bhujuk; abum seke 220,
" chepuk.

" 225)26(0   0   Raas 660

   26)225(8   8   5720
     208
                 1   26)660(25
      17)26(1
       17   1   10 seke the lubd,
        9)17(1   1
        9   225)5720(25
          220
         8)9(1   95 seke the goonuk.
         8   0

       1 disregarded.

$$\frac{26 \times 95 - 220}{225} = 10$$

" Hence the chandra days are 95."

The last rule of this chapter is taken notice of as follows :

" OF THE SANSTIST COOTUK."

" By what number may 5 be multiplied and divided by 63, the remainder will
" be 7 ; and that number so found, when multiplied by 10 and divided by 63,
" the remainder will be 14.
" The two goonuks are 5 and 10, the sum is the bhady ; the two sekes are
" 7 and 14, the sum is the chepuk ; the bhujuk in both is the same or 63."

The question is solved as before ; it ends " Thus, however numerous be
" the goonuks given, let them all be added for the bhady ; and the same with
" respect to the given sekes for the chepuk ; the bhujuk will be always the
" same."

Chapter 6.

" THE CHACRA BALA.

—••••◦|◦◦••—

"**T**HE multiplication of the square is a chacra bala. There are six cases:
" The first quantity assumed is called *hursua* (the smaller); its square must be
" multiplied by the *pracrit*, and then must be added the *chepuk*; that is such a
" chepuk as will by addition produce a square, and this chepuk may require to be
" affirmative or negative, which must be ascertained. The root of this square is
" the *jeist*: these three, the canist or hursua, jeist, and chepuk must be noted
" down and again written down."

The distinctions of *samans babna* and *anter babna* are given as follows:

" OF THE SAMANS BABNA."

" When the jeist and canist are multiplied into each other (budjra beas)
" the sum is the hursa or canist. It is called *budjra beas* from its being a tri-
" angular multiplication; the upper, or jeist, or greater, being multiplied by the
" lower, smaller, the canist; and the canist multiplied by the greater or jeist;
" the two products added is the hurs.

" The two canists multiplied together, and multiplied again by the pracrit,
" then the product of the two jeists—added altogether, produces the root of the
" jeist; the product of the two chepuks then becomes the chepuk."

The anter babna is described thus: " The difference between the two products
" or budjra beas, produces hursa or canist. The product of the canists multiply
" by the pracrit, and the difference between (*this and*) the product of the two
" jeists is the root of the jeist, and the product of the two chepuks is the
" chepuk."

The rest of this is very imperfect, but the cases of $\beta\beta = Bp^2$ and $\beta\beta = \dfrac{B}{p^2}$, and

the rule $A \left(\dfrac{2r}{r^2 - A} \right)^2 + 1 = \square$, are plainly alluded to. (See notes on the Persian

translation.) " Thus" (it is added) " the root of the canist and jeist may be in a
" variety of cases found."

After this there are examples the same as in the Persian translation, and worked
the same way as far as the " Operation of Circulation ;" and, after the examples.
" Hence, how various soever the Ist, from the somans babna and anter babna may
" be produced canist, jeist, and chepe; and hence it is called the chacra bala."

I find no abstract of the rule for the " operation of circulation," but there is
the first example, viz. $67x^2 + 1 = \square$, as follows : " Roop 1 is the canist, $\dot{3}$ is
" the chepe; then the pracrit 67, canist 1, jeist 8.   Hurs 1 is the bhady, chepe is
" the bhujuk, jeist 8 is the chepuk ; then by the cootuk gunnit

" Bha. 1, che 8,

" Hur $\dot{3}$; hence the goonuk 1.

" then the square of 1 is 1, $67 - 1 = 66$, but this is not the smallest; then
" $\dot{3}+\dot{3}=6$, $6+1=7$; its square 49, deduct from pracrit $67-49=18$; 3)18($\dot{6}$;
" but the negative must be made affirmative 6; and $5\times5=25$, and $25\times67=$
" 1675, and $1675+6=1681$ its root 41 ; then by the cootuk gunnit

" Bha 5, che 41,

" Hur 6 ;

" then $5\times5=25$ and $61-25=42$, $\dot{6}$)42($\dot{7}$; the lubd is the canist 11 ; $11\times11=$
" 121, $121\times67=8107$; chepe is $\dot{7}$, $8107+\dot{7}=8100$ its root is 90, which is the
" jeist ; then by the cootuk

" Canist is bhady 11, che 90,

" Hur $\dot{7}$.

" Here the goon is 2, che $\dot{7}$; $7+2=9$ the second goonuk ; its square is 81
" $81-67=14$; 7)14(2 the other chepe."

" The canist 27.   This is made jeist 221.

" Ca. 27, 221 jeist, che 2.
" Ca. 27, 221 jeist, che 2.

" Ca. 11934, jeist 97684, che 4.
" Ca. 5967, jeist 48842, che 1.

" The square ca. 35605089, which multiply by 67, and 1 added, the sum will
" be 2385540964, and its root is 48842."

BOOK 1.

———◆———

"**T**HE unknown quantities, &c. must be clearly stated, and then the equation
"must be reduced in the manner hereafter shewn by ✕, by ÷, by the rule of
"proportion, by progression, ratios, by △ ; still maintaining the equality.
"When they are otherwise, add the difference; then *sodana* the quantities; the
"same with respect to roots. In the other side of the equation the roop must be
"sodanad with the roop. When there are surds they must be sodanad with
"surds; then by the remainder of the unknown quantities division, the roop
"must be divided; the quotient is the quantity sought, now become *visible*."

"Then the quantity so found must be utapanad, in order to resolve the
"question."

It will be remarked that the Persian translation has "*the figure of the bride*,"
for that expression which is represented by △ in the above abstract. Mr.
Davis tells me that the original had nothing like a reference to Euclid, and that
this part related simply to the proportions of right-angled triangles.

There follow abstracts of the seven first questions of this book, with their
solutions, which are the same as those in the Persian translation.

The first part of the first example is: "One man had 6 horses and 300 pieces of
"silver, and the other had 10 horses, and owed 100 pieces of silver; their pro-
"perty was equal. *Quære*, the value of each horse, and the amount of the pro-
"perty of each person. Here the unknown quantity is the price of one horse.

"Ja 6, roo 300
"Ja 6, roo 100 these are equal.
———————————
"Ja 6, roo 300
"Ja 10, roo 100. Sodan, that is transpose.

"Ja 6 + 300 = Ja 10 − 100
"Ja 4 = 400
"Ja   = 100

The third example, where the Persian translator has introduced the names Zeid and Omar, is in Mr. Davis's notes thus:

" One man said to another, if you give me 100 pieces of silver I shall have " twice as many as you ; the other said give me 10 pieces and I shall have six " times as many as you. *Quære*, the number each had.

" Ja 2 roo 100

" Ja 1 roo 100

" Ja 12 roo 660

" Ja 1 roo 110

" Diff. Ja 11 roo 770

" Ja   roo 70

BOOK 2.

" **T**HE square root of the sum of the squares of the bhoje and cote is the carna. " Explain the reason of this truth.

" The carna is ka ja ; the figure thus, Divide this by a perpen-

" dicular ; these are equal triangles. The bhoje is abada or given.

" The lumb or perpendicular is the cote, In the latter the cote

" is a carna, the lumb perpendicular is the bhoje, the cote is the carna ; they are " similar triangles, When the bhoje, now carna, gives the lumb for the cote, " then cote for carna how much ? Thus by proportion the cote is found." Also

" As bhoje 15 is to carna, then from this carna 15 what bhoje ?

o

" Therefore 15 × 15, and divide by Ja 1, and the small bhoje is found =
$\frac{225}{Ja\ 1}$. Again.

" As cote 20, to the carna, so is the carna 20. What cote?"

This is found = $\frac{400}{Ja\ 1}$, and this added to $\frac{225}{Ja\ 1} = \frac{625}{Ja\ 1}$ is the carna; whence
25 = Ja 1.

" Then from the bhoje to find the perpendicular.

" The bhoje 15, its square 225; bhoje abada = 9, its square 81; the differ-
" ence is 144; its root is the lumb 12.

" So, the cote 20, its square 400; cote abada 16, its square 256; difference
" of squares 144; its root the lumb is 12." Again,

Another way.

" Carna ja 1; then half the rectangle of the bhoje and cote is equal to the
" area = 150; therefore the area of the square formed upon the carna in this
" manner will be equal to four times the above added to the contained square,
" which square is equal to the rectangle of the difference between the bhoje and
" cote, which is 5 × 5 = 25. The rectangle of the bhoje and cote is 15 × 20
" =300; and 300×2=600 (or $\frac{300}{2}$ × 4); 600 + 25 = 625, which is equal to
" the area of the whole square drawn upon the carna, and therefore the square
" root of this is equal to the carna = 25. If this comes not out an integral
" number, then the carna is imperfect or a surd root.

" The sum of the squares of the bhoje and cote, and the square of the sum of
" the bhoje and cote, the difference of these is equal to twice their rectangles;
" therefore (theorem) the square root of the squares of the bhoje and cote is equal
" to the carna. To illustrate this, view the figure."

Here a figure is given which requires explanation to make it intelligible.

" In that figure where 3 deducted from the bhoje, and the square root made of
" the remainder, and one deducted from the square root, and where the remainder
" is equal to the difference between the cote and carna. Required the bhoje, cote,
" and carna.

" OPERATION."

" Let the assumed number be 2, to which add 1, its square is made= 9; to this
" add 3, whence the bhoje is 12; its square is 144, and this by the foregoing is

" equal to the difference between the squares of the cote and carna ; and the sum
" of the cote and carna multiplied by their difference is equal to this."

Then follows something which I cannot make out, but it appears to be an
illustration of the rule, that the difference of two squares is equal to a rectangle
under the sum and difference of their sides, probably the same as that in the
Persian translation. The end of it is,

" Thus the square of 5 is 25, and the difference between 5 and 7, sides of the
" square, is 2 ; the sum of those sides is 12, which multiplied together is 24 ;
" therefore equal to this is the remainder, when from the square of 7 is deducted
" the square of 5.

" The difference between the squares of these is known, and thence the
" cote and carna are discovered thus : This difference of squares divide by the
" difference of the cote and carna, or difference of roots, as in the Pati Ganita,
" $\frac{144}{2} = 72$, and this is the sum of the two quantities sought, as is taught in the
" Pati Ganita, but their difference is 2 ; therefore deduct 2 from the sum, the
" remainder is 70, and half of this is the first quantity sought. Again, add 2 to
" 72, the sum is 74 ; its half is 37 the other quantity ; therefore the cote is 35,
" the carna 37.

" When the proposed difference is 1, the numbers are found 7, 24, 25 ; multi-
" ply these by 4, the numbers will be 28, 96, 100.

Then follows a note of the rule, that the difference of the sum of the squares of
two numbers, and the square of their sum, is equal to twice the rectangle of the
two numbers, and this example as in the Persian translation.

" The two numbers are 3 and 5 ; the sum of squares $9 + 25 = 34$; the sum
" 8, its square 64 ; the difference is $64 - 34 = 30$; then $5 \times 3 = 15$, $15 \times 2 = 30$,
" equal to the above. But when the sides are not known, but the difference of
" their squares, 16 then divide by 2, (viz. *by the difference of the numbers)*
" $\frac{16}{2} = 8$; this is their sum, and deduct their difference $8 - 2 = 6$, half this is

" one number, and $8 + 2 = 10$, and $\frac{10}{2} = 5$, the other number."

The next is,

" In the figure where the sum of bhoje, cote, carna is 40, and the product of
" bhoje and cote 120. What is the bhoje, cote, carna?

" Multiply the product 120 by 2 = 240, this will be equal to the difference between
" the square of the sum of the bhoje and cote and the carnas square. The sum
" of the squares of the bhoje and cote equal to square of the carna; therefore the
" product of the bhoje and cote × by 2 is equal to the difference between the
" rectangle and cote (the square) of the sum of the bhoje, and the square of the
" carna.

" Divide this number 240 by the sum of the bhoje, cote, and carna 40, $\frac{240}{40} = 6$,

" which is equal to the difference between the carna and the sum of bhoje and

" cote. Hence $\frac{40 - 6}{2} = 17$ the carna : 23 sum of bhoje and cote, squared is

" 529. Multiply the rectangle of bhoje and cote 120 by 4 = 480, the remainder
" 49, and its root 7 ; this is the difference of bhoje and cote ; deduct this from

" their sum 23 ; 23 − 7 = 16, its half $\frac{16}{2} = 8$ is the bhoje ; 23 + 7 = 30, its

" half, is the cote 15."

The next is,

" Where the sum of bhoje, cote, carna is 56, and their product 4200, what
" are the bhoje, cote, carna?

" Ja 1, ja, bha 1. The sum of bhoje, cote, carna.

" Carna ja 1 ; ja 1, roo 56 ; these three multiplied, 4200.

" The rectangle of bhoje and cote $\frac{roo\ 4200}{ja\ 1}$ equal to sum of squares of bhoje

" and cote is ja bha 1, sum of bhoje and cote ja 1, roo 56 ; the square ja bha 1,

" ja 112, roo 3136 ; the difference between them is equal to $\frac{8400}{ja\ 1}$;

" therefore
Ja 112	roo 3136
Ja 0	roo 8400

ja 1

" divide both by 112; reduce both sides, and it will be

$$\frac{\text{Ja 1} \quad \text{roo 28}}{\text{Ja 0} \quad \text{roo 75}}$$
$$\text{Ja 1}$$

Reduce the fractions.

$$\frac{\text{Jȧ bha 1,} \quad \text{ja 28,} \quad \text{roo} \quad 0}{\text{Ja bha 0,} \quad \text{ja} \quad 0, \quad \text{roo 75}}$$

" Multiply by 4, and add the square of 28.

$$\frac{\text{Ja bha 4,} \quad \text{Jȧ 112,} \quad \text{roo 300} \; (\textit{should be} \; 784)}{\text{Ja bha 0,} \quad \text{Ja 0} \quad \text{roo 484}}$$

" The square root $\frac{\text{ja 2} \quad \text{roo 28}}{\text{ja 0} \quad \text{roo 22}}$; then add, $\frac{50}{\text{ja 2}}$; divide by $2 = 25$, which is
" the jabut, and therefore carna.

" Then for the bhoje cote. The three multiplied are 4200. Divide by carna
"$\frac{4200}{25} = 168 = $ bhoje $\times$ by cote. The sum of bhoje and cote $= 56 - 25 = 31$,
" and $168 \times 4 = 672$. The square of $31 = 961$," (the difference) " 289, its square
" root is the difference of bhoje and cote $= 17$; deduct this, $31 - 17 = 14$; its
" half 7, which is the bhoje; and $31 + 17 = 48$; its half 24 is the cote."

The lines above have been carelessly drawn. The true Hindoo method of
writing the equation $- x + 28 = \frac{75}{x}$ I understand to be this, $\frac{\text{ja 1} \mid \text{roo 28,}}{\text{Ja 0} \quad \text{Roo 75}}$ and
$$\text{Ja 1}$$

that of $- x^2 + 28x = 75$ this, $\frac{\text{Ja bha 1} \mid \text{ja 28} \mid \text{roo} \quad 0}{\text{Ja bha 0} \mid \text{ja} \quad 0 \mid \text{roo 75}}$

Books 3, 4, and 5.

————••◦◦◦◦◦••————

I FIND among Mr. Davis's notes a small part only of the beginning of the 3d book, which consists of rules for the application of the cootuk to questions where there are more unknown quantities than conditions. I find also some notes which evidently relate to the first example of this book, but nothing distinct can be made out.

There are no notes relating to the 4th book.

Of the 5th book only this:

" When there are two or more quantities multiplied, the 1st quantity must be " discarded—then"....There is also an abstract of the first example, the same as that in the Persian translation.

————◄◄◄◄◄◄◄◄◄◄◆►►►►►►►►————

Extracts from Mr. Davis's Notes, taken from a modern Hindoo Treatise on Astronomy.

————••❊••————

" **By** the method of the Jeisht and Canist from two jyas* being found, others " may be computed by those who understand the nature of the circle (the bow

---

* *Jya* or *jaw*; sine.—The modern Europeans acquired their knowledge of the *sine* from the Arabians; and it is obvious that they used the term sinus only, because the word *jeeb* (جيب), by which the Arabians called the line in question, is translated *sinus indusii*. The radical meaning of (جيب) is to cut, and it denotes the bosom of a garment only, because the garment is cut there to make a pocket; accordingly we find that جيب

" and arrow), and thus, by the addition of surds, may the sum and the
" difference of the arc and its sine be computed whether that arc be 90 degrees,
" more or less.

---

does not mean bosom, but that among the Arabians it signifies that part of their dress where the pocket is usually placed, and in some languages which abound in Arabic words, as the Persian and the Hindoostanee, it is the common term, not only for a pocket in the bosom, but for any pocket wherever it may be. In all Arabic dictionaries this word is explained as above, and in some, though not in all, (it is not in the Kushfool Loghat) the line we call *sine* is given as a second meaning.

The Arabs call the arc *kous* (قوس), which signifies *a bow*; the cord *wutr* (وتر), which is the *bow-string*; and the versed sine *suhum* (سهم), which is the *arrow*. But the sine they express by a word which has no connexion whatever with the *bow*.

The Mathematical history of the Arabians is not known enough for us to speak positively about the first use of sines among them, but there seems to be reason to suspect that they had it from a foreign source, probably from the Indians.

The Sanscrit word for the chord is *jaw*, or more properly *jya* and *jiva*. (For these terms see Mr. Davis's paper in the second volume of the Asiatic Researches; the literal explanation of the words has been given me by Mr. Wilkins,) and the sine is called *jya ardhi*, or half cord; but commonly the Hindoos, for brevity, use *jya* for the sine. They also apply the word in composition as we do; thus, they call the cosine *cotijya*, meaning the *sine*, the side of a right-angled triangle; the sine (or right sine) *bhojjya*, meaning the *sine*, the base of a right-angled triangle, and *cramajya* the *sine* moved; the versed sine they call *ootcramajya*, or the *sine* moved upwards; the radius they call *tridjya*, or the *sine* of three, (meaning probably three signs.) In their term for the diameter *jyapinda*, or whole *jya*, the word is used in its proper acceptation for chord, and not for *jya ardhi*, or sine.

It seems as if جيب and *jya* were originally the same word. Mr. Wilkins (the best authority) assures me that *jya*, in the feminine *jiva*, is undoubtedly pure Sanscrit, that it is found in the best and oldest dictionaries, and that its meaning is a *bow-string*.

The Arabians in adapting a term to the idea of chord, had reference to the thing which it resembled, and called it وتر or the *bow-string*; but having so applied this term, they had to seek another for *sine*; then they would naturally refer to the name of the thing, and call it by some word in their own language, which nearly resembled that under which it was originally known to them. This mode of giving a separate designation to the sine was evidently more convenient than that of the Hindoos, so I conjecture that جيب for sine is no other than the Sanscrit word *jya* or *jiva*.

It is remarkable that the Sanscrit terms for the sides of a right-angled triangle have reference to a *bow*: they seem to be named from the angular points which are formed by the end of the bow, the arm which holds it, and the ear to which the string is drawn; thus the side is called *coti*, or *end of the bow*; the base *bhoj*, or the arm; and the hypothenuse *carna*, or the *ear*. Some further explanation however is desirable to shew why *bhojjya* is the term for the sine, and not (as it should be by analogy) the cosine, and *cotijya* the cosine instead of the sine.

The Hindoos have a word for the versed sine, *sur*, which signifies arrow, answering exactly to the Arabic سهم

" Multiply the jaw of one of two arcs by the cotejaw of the other arc, divide
" the product by the tridjaw, add the two quotients and also subtract them ; the
" sum is equal to the jaw of the two arcs, the other is the jaw of the difference
between the two arcs.

" Again, multiply the two bojejaws together, and likewise the two cotejaws
" together ; divide by the tridjaw. Note the sum and the difference. The sum
" is the cotejaw of the sum of the two arcs, the difference is the cotejaw of the
" difference of the two arcs.

" In this manner Bhascara computed the sines in his Siromony, and others
" have given other methods of their own for computing the same.

The author of the Marichi observes, " that the author of the Siromoni derived
" his method of computing his sines by the jeisht and canist, and diagonally multi-
" plied (ba jera beas), the jeisht and canist being the cotejaw and the bojejaw;
" hence he found the sines of the sum and difference of two arcs, the third
" canist being those quantities. He did not use the terms jeisht and canist, but
" in their room bojejaw and cotejaw. I shall therefore explain how they
" were used.

" The bojejaw = canist (small).

" Cotejaw = jeisht (larger).

" (The theorem then is what square multiplied by 8, and 1 added, will produce
" a square).

" Multiply the given number (8) by the square of the canist, and add the
" chepuk, the sum must be a square.

" The bojejaw square deducted from the tridjaw square, leaves the cotejaw
" square, therefore the bojejaw square is made negative, and the tridjaw square
" added to a negative being a subtraction, the tridjaw square is made the chepuk.

" The canist square, which is the bojejaw square, being multiplied by a negative
" becomes a negative product, therefore the quantity is expressed by 1 roop
" negative.

" Then the bojejaw square multiplied by 1 roop negative, and added to the
" tridjaw, its square is the cotejaw.

" Hence the bojejaw and cotejaw in the theorem by Bhascara, represent the
" canist and jeisht, and 1 roop negative is the multiplier, and the chepuk is the
" square of the tridjaw, and the equation will stand as follows :

" Canist 1st. jaw 1 : jeisht 1st. cotejaw 1 : chepuk, tridjaw square 1,

" Canist 2d. jaw 1 : jeisht 2d. cotejaw 1 : chepuk, tridjaw square 1.

" These multiplied diagonally produce
  " 1st jaw 1. 2d cotejaw 1.
  " 2d jaw 1. 1st cotejaw 1.

" These added produce the first canist, viz.
  " 1st jaw + 2 cotejaw.
  " 2d jaw + 1 cotejaw,
" which is the sum (or joge) and the difference.

  " 1st jaw — 2 cotejaw.
  " 2d jaw — 1 cotejaw.

" Thus from the sum and difference are produced two canists, and the square of
" the tridjaw squared is the chepuk; but the chepuk wanted being only the
" square of the tridjaw, then as the Bija Ganita directs divide by such a number
" as will quote the given chepuk.

" Therefore the tridjaw being the ist, or assumed, or given quantity, divide
" the canist by it, the quotient will be the tridjaw square, and hence the theorem
" in Bhascara for the bojejaw.

" And in like manner the cotejaws are found; but Bhascara did not give this
" theorem for the cotejaws, because it was more troublesome. He therefore gave
" a shorter rule. But since the cotejaw square is equal to the bojejaw square
" deducted from the tridjaw, therefore the same rule may be applied to the
" cotejaw, by making the cotejaw the canist, and the bojejaw jeisht; then by the
" foregoing rule the cotejaw of the sum or difference of the arcs may be found".

Second Extract.

"SLOCA. The munis determined the equations of the planets centres for the " use of mortals, and this can be effected only by computations of the sines of " arcs. I shall explain and demonstrate their construction and use.

" 2. And for this purpose begin with squares and extractions of roots, for the " satisfaction of intelligent persons of ready comprehension.

" 3. The square is explained by the ancients to be the product of a number " multiplied by itself. (He goes on to show how squares are found and roots " extracted as in the Lilavati).

" 6. Square numbers may be stated infinitely. The roots may be as above " extracted, but there are numbers whose roots are irrational. (Surds.)

" 7. The ancients have shewn how to approximate to the roots of such " numbers as follows: Take a greater number than that whose root is wanted ; " and by its square multiply the given number, when that given number is an " integer. Extract the root of the product, divide this root by the assumed " number, and the quotient will approximate to the root required. If the given " number be a fraction, multiply and extract as before. To approximate the " nearer the munis assumed a large number, but the approximation may be made " by assuming a small number."

And after a blank.
" In like manner surds are managed in the *abckt* or symbolical letters, " (Algebra) expressing unknown quantities."

Again, after a blank.
" Some have pretended to have found the root of a surd, and that this might

" be effected by the Cutuca Ganita, attend and learn whether or not this could
" have been possible.   I shall relate what Bhascara and others have omitted to
" explain.   A root is of two kinds; one a line, the other a number.   And the
" root of a square formed by a line expressing 5, may be found, though the root
" of 5 cannot be numerically expressed ; but the numbers 1, 4, 9, &c. may be
" expressed both ways.   2, 3, 5, &c. are surds, and can have their roots expressed
" only by lines.   (He goes on to shew the impossibility of finding the root of a
" surd, though it should be eternally pursued through fractional quantities.)
   " The root of a surd may be shewn *geometrically.*"

I Have copied these two extracts exactly as I found them ; there appear to be
one or two errors which it may be as well to mention.   In the first extract the
latter part of the first sentence should, perhaps, run thus : " By the addition of
" the jeisht and canist may the sines of the sum and difference of arcs be
" computed," &c.

I observe that where jeisht and canist first occurred in these notes Mr. Davis
translated it originally " arithmetic of surds," and afterwards corrected it ;
probably from oversight it was not corrected in the second place,

The value of the cosine of the sum of two arcs is given instead of that of the
difference and *vice versa.*

There is an error also in writing the sum and the difference of the cross products.

I know nothing of the author of the Marichi.   Possibly he might have
observed that the jeisht and canist rule corresponded with the formulæ for the sines
and cosines, and the latter were not derived from the former by Bhascara, but
invented at a later period, or introduced among the Hindoos from foreign
sources.   Probably however the application and the formulæ are both of Indian
origin.

As for the second extract the rule for approximating to the square root is the
same as that given by Recorde, in his " Whetstone of Wit," which was published in 1557; and by his contemporary Buckley; (for an account of whose
method see Wallis's Algebra, p. 32. English edition.)   I have before stated, that

this rule is also in the Lilavati.   I mentioned it generally then only because of its connexion with a trigonometrical proposition.   The following is a literal translation of the rule, as given by Fyzee: " Take the squares of the base and " side, and add them together; then multiply by the denominator and write it " down.   Then assume a large number and take its square.   Then multiply it " by that which was written down.   Take the square root of the result and call " it the dividend.   Then multiply that denominator by that assumed number, " and call it the divisor.   Divide the dividend by the divisor, the quotient is the " hypothenuse."   This is not delivered with perfect accuracy, the true meaning however is plain.   If the assumed multiplier is decimal the method gives the common approximation in decimal fractions.   The writer denies that the root of a surd can be found by the cootuk, but he speaks of it as a subject to which the cootuk was said to have been applied.   It is very improbable that such a thing as this should have found its way from Europe to India, and it is very probable that many things of this sort were to be had from Hindoo sources.

Explanation of Sanscrit Words used in Mr. Davis's Notes.

———◆———

BIJA GANITA—Algebra—Literally seed counting.

Pati Ganita—Arithmetic—Ganita seems to be used as we use arithmetic. Thus as we have arithmetic of integers, arithmetic of surds, decimal arithmetic, &c. the Indians have bija ganita, pati ganita, cutuca ganita, &c.

Jabut Tabut—The unknown quantity, as we use x—Literally *as far, so far.* It is not clear how this comes to be so used. It would be more conformable to the rest of Hindoo notation, if the word pandu (white) were applied; the first letter of pandu is very like that of jabut, and they might easily be confounded.

Kaluk, neeluk, &c.—Unknown quantities—Literally the colours black, blue, &c.

Abekt—Unknown,

Carni, surd—Hypothenuse—Literally ear.

Mahti and laghoo—Greater and less.

Roop—Known quantity—Literally form, appearance.

Bhady—Dividend.

Hur—Divisor.

Bhujuk—Divisor.

Seke—Remainder.

Dirl—Reduced.

Chepuk or *chepe*—Augment.

Bullee—Chain or series.

Unte—Last.

Upantea—Last but one.

Lubd—Quotient.

Goonuk or *Goon*—Multiplicand.

Rhin—Minus—Literally decrease.

Coodin—An astronomical period.

Urgan—Number of days elapsed.

Bekullas—Seconds.

Cullas—Minutes.

Ansas—Degrees.

Bhaganas.—Revolutions.

Calp—The great period.

Raas—Literally a heap, a sum total, a constellation.

Cootuk—The principle on which problems of this form $\dfrac{ax + c}{b} = y$ are solved.

Sanstist—Ditto of $\dfrac{ax}{b} = y + c$ and $\dfrac{dx}{b} = z + e.$

Chacra-bala—Ditto of $ax^2 + b = y^2$—Literally strength.

Hursua, hurs, hursa—x in the above form—Literally small.

Pracrit—a in Ditto—Literally principal.

Jeist or *Jeisht*—y in Ditto—Literally greatest.

Canist—x in Ditto—Literally least.

Samans babna—If $Ax'^2 + \beta = y'^2$ and $Af^2 + \beta = g^2$, then the rule $x'' = x\,g + y'f$ is called samans babna—Literally contemplation of equal degrees.

Anter babna—In the above form, when $x'' = x'g - y'f$, it is called anter babna—Literally contemplation of difference.

Badjra beas—Cross multiplication which produces the above forms—Literally cross diameter.

Cootuk gunnit or *cutuca ganita.*—Cootuk Calculation.

Sodana—Reduce—Literally purify.

Utapana—Brought out.

Bhoje—Base of a right-angled triangle—Literally arm.

Cote—Side of Ditto—Literally end of a bow.

Carna—Hypothenuse—Literally ear.

Lumb—Perpendicular—Literally length.

Abada—Given.

Ist—Assumed.

Jaw or *Jya*—Sine or chord—Literally bow-string.

Bojejaw—Sine.

Cotejaw—Cosine.

Tridjaw—Radius—Literally sine of three; perhaps meaning of three signs or 90 degrees.

Addy—Intercalary.

Che-tits (Cshaya tithi)—Difference of solar and lunar days.

Abum—? For bhumi savan—solar days.

Chandra—Lunar.

For the literal explanation of these terms, as far as they could be made out, I am obliged to Mr. Wilkins. Most of the words are written here according to their common pronunciation in Bengal.

DEAR STRACHEY,

HAVING just laid my hands on a parcel of papers of notes, containing abstracts and translations from the Bija Ganita, made by me, with the assistance of a Pandit, as long ago as when I was stationed at Bhagulpore*, I send them to you with full liberty to make any use of them. Ever since my removal to Burdwan these papers have lain unnoticed, and might have continued so had it not occurred to me that you are occupied in such researches. There may be trifling inaccuracies in some places, the translations having been made carelessly and never revised; but their authenticity may be depended on, as they were made from the original Sanscrit Bija Ganita, which was procured for me at Benares, by Mr. Duncan. I send also a book of memoranda, containing chiefly trigonometrical extracts from a modern astronomical work in Sanscrit, which I suppose to have been written in Jey Sings time.

I am very sincerely your's,

Portland Place, Jan. 1812.   S. DAVIS.

THE END.

---

* About the year 1790.

Glendinning, Printer, Hatton Garden, London.